Calculus

DENISE SZECSEI

CAREER PRESS
Franklin Lakes, NJ

Homework Helpers: Calculus
Edited by Jodi Brandon
Typeset by Eileen Dow Munson
Cover design by Lu Rossman/Digi Dog Design NYC
Printed in the U.S.A. by Book-mart Press

To order this title, please call toll-free 1-800-CAREER-1 (NJ and Canada: 201-848-0310) to order using VISA or MasterCard, or for further information on books from Career Press.

The Career Press, Inc., 3 Tice Road, PO Box 687,
Franklin Lakes, NJ 07417
www.careerpress.com

Library of Congress Cataloging-in-Publication Data

Szecsei, Denise.\
Homework helpers. Calculus / by Denise Szecsei.
p. cm.
Includes index.
ISBN-13: 978-1-56414-914-5
ISBN-10: 1-546414-914-5
1. Calculus. 2. Calculus—Problems, exercises, etc. I. Title.

Qa300.S985 2006
515—dc22

2006026398

Dedication

This book is dedicated to Mickey Perry; the first of my guides through the 9 circles of mathematics.

Acknowledgments

This book was a group effort, and I would like to thank the people who helped throughout the entire production.

I would like to thank Michael Pye, Kristen Parkes, and everyone else at Career Press who worked on this project. Jessica Faust handled the logistics so that I could focus on writing.

I am grateful for Kendelyn Michaels's willingness to assist in the development of this book. I benefited greatly from her review of the manuscript and her suggestions for improvements. I appreciate her continued participation in my writing projects.

Alic Szecsei helped reduce the number of typographical errors in the manuscript and was willing to spend many hours working out the solutions to the review problems.

I want to thank my family for their understanding and patience throughout the writing process. I could not have completed this project without their support.

Contents

Preface

Welcome to *Homework Helpers: Calculus*!

Calculus is a tool that can be used to analyze functions. Two mathematicians, Sir Isaac Newton and Gottfried Leibniz, are credited with developing calculus in the 17th century. There is some controversy regarding the development of calculus. Leibniz published his results before Newton did, and Newton claimed to have developed calculus years earlier but delayed publication. Newton took things a step further and claimed that Leibniz's work was based on ideas taken from notes and letters written *by* Newton. It's hard to imagine *anyone* fighting over a mathematical result, but politics permeates even the field of mathematics.

Calculus involves the study of limits. The Greeks used limits, or the method of exhaustion, to approximate pi and to compute the area of a region and the volume of a solid. The application of calculus to computing areas and volumes is called *integral calculus*.

Calculus is also the study of change. Newton's motivation for developing calculus was to study motion and rates of change as they apply in physics. Calculus can be used to describe Newton's laws of motion, and Newton's notation is widely used in physics. The application of calculus to motion and rates of change is called *differential calculus*.

Calculus marks the transition from working with the *static* nature of a function to analyzing the *dynamic* nature of a function. We will be moving away from calculating the value of a function at a particular point and moving towards developing an understanding of how a function changes over a particular interval, or over time.

The ideas in calculus can be explained using everyday language. Calculus problems usually involve one or two steps that actually fall under the realm of calculus. Most of the difficulty in solving calculus problems is algebraic in nature. Calculus problems involve a lot of factoring, multiplying, and dividing polynomials. Solving calculus problems usually requires solving exponential, logarithmic, and trigonometric equations. Calculus will *build* on the algebraic skills used to analyze a function, not *replace* them. Because of this, I will spend the first few chapters reviewing the properties of functions from an algebraic perspective. Laying a strong algebraic foundation will definitely pay off in our calculus explorations.

The skills that you will develop in the process of learning calculus can be applied to almost every other field of study imaginable. Physics, chemistry, biology, business, economics, sociology, and medicine are just a few areas where your knowledge of calculus and advanced mathematical problem-solving skills will enable you to excel.

One of the most interesting ideas in calculus has to do with existence theorems. An existence theorem is a claim that an object exists without actually producing the object. The main existence theorems in calculus are the Intermediate Value Theorem, the Extreme Value Theorem, Rolle's Theorem, and the Mean Value Theorem. Though these theorems are very important results on their own, they will also provide you with some insight into what mathematicians do: We hypothesize, conjecture, and prove theorems. The other important theorem we will discuss is the Fundamental Theorem of Calculus. This theorem bridges the gap between differential calculus and integral calculus.

I wrote this book with the hope that it will help anyone who is struggling to understand calculus or is just curious about the subject. Reading a math book can be a challenge, but I tried to use everyday language to explain the concepts being discussed. Looking at solutions to math problems can sometimes be confusing, so I tried to explain each of the steps I used to get from Point A to Point B. Keep in mind that learning calculus is not a spectator sport. In this book I have worked out many examples, and I have supplied practice problems at the end of most of the lessons. Work these problems out on your own as they come up, and check your answers against the solutions at the end of each chapter. Aside from any typographical errors on my part, our answers should match.

I hope that in reading this book you will develop an appreciation for the subject of calculus and the field of mathematics!

A Review of Functions

The concept of a function is very important in mathematics. Functions can be used to describe, or model, many situations in our everyday lives. In economics, functions can be used to calculate income tax, interest earned from an investment, and monthly loan payments. In science, functions can be used to predict the pressure exerted by a gas, the energy released in a chemical reaction, and the occurrence of the next lunar eclipse. The study of calculus requires a solid understanding of some basic elementary functions. It is common for problems in calculus to involve only one or two steps that actually pertain directly to calculus! The majority of the work in solving problems in calculus is algebraic in nature and involves analyzing functions. This chapter should not only serve as a review of the general properties of functions, but also help you understand why it is very important to have a strong foundation in algebra.

Lesson 1-1: Representing Functions

A **function** is a set of instructions that establishes a relationship between two quantities. A function has input and output values. The input is called the **domain** and the output is called the **range.** The variable used to describe the elements in the domain is called the **independent variable.** The variable used to describe the output is called the **dependent variable**, as it depends on the input. An important feature of a function is that every input value has only one corresponding output value. A function can be represented in a variety of ways. Functions can be described using words, a formula, a table, or a graph.

Using a formula to define a function is a convenient way to describe the function in mathematical terms instead of using words. When analyzing a formula, it is important to use the order of operations. A function is often written $y = f(x)$, where $f(x)$ is an algebraic expression that involves the independent variable x. The variable x is sometimes referred to as the **argument** of the function. Evaluating the function for a particular value of x involves replacing every instance of the independent variable with that value.

Scientists often collect data from various instruments and record this information in a table. These tables represent functions, and they provide an easy way to describe a complex, or unknown, formula.

The graph of a function is usually presented using the **Cartesian coordinate system.** In the Cartesian coordinate system, we use a vertical line, called the **y-axis**, and a horizontal line, called the **x-axis**, to divide the plane into four regions, or **quadrants.** The intersection of these two lines is called the **origin.**

Two numbers are used to describe the location of a point in the plane, and they are recorded in the form of an ordered pair (x, y). A function can then be thought of as a collection of ordered pairs (x, y). The graph of a function is then the graph of these ordered pairs on the coordinate plane.

If the graph of the function $f(x)$ crosses the x-axis, then we say that $f(x)$ has an x-intercept. The **x-intercept** of $f(x)$ is the point where $f(x)$ crosses the x-axis. Because any point on the x-axis has a y-coordinate of 0, the x-intercept of $f(x)$ corresponds to a point of the form $(a, 0)$. The x-intercepts of a function are often called the **roots** of the function, or the **zeros** of the function. Not all functions have x-intercepts. Finding the x-intercepts of a function involves solving the equation $f(x) = 0$ for x.

If $x = 0$ is in the domain of a function $f(x)$, then the point $(0, f(0))$ is the y-intercept of $f(x)$. The **y-intercept** of a function represents where the function crosses the y-axis. Not all functions have y-intercepts, but if a function has a y-intercept, the y-intercept will be unique. In other words, a function can have at most one y-intercept.

Lesson 1-2: The Domain of a Function

The **domain** of a function represents the allowed values of the independent variable. If a function is described using words, then the domain needs to incorporate the context of the description of the function. For example, if a function describes the number of buses needed for a field trip as a function of the number of expected passengers, then the domain of this function cannot include any negative numbers: Transporting a negative number of passengers makes no sense!

The description of a function using a formula may or may not include a domain. If the domain is not indicated, then it is safe to assume that the domain is the set of all real numbers that, when substituted in for the independent variable, produce real values for the dependent variable. To find the domain of a function, start with the set of all real numbers and whittle down the list. There are two things that are frowned on in the mathematical community. The first thing that is forbidden is to divide a non-zero number by 0; quotients such as $\frac{3}{0}$ are meaningless. The second thing that is not allowed in the world of real numbers is to take an even root of a negative number; for example, there is no real number that corresponds to $\sqrt{-4}$. To find the domain of a function that involves an even root, such as a square root, a fourth root, and so on, set whatever is under the root to be greater than or equal to 0 and find the solutions to the inequality. Then toss out any points that result in the denominator being equal to 0.

Example 1

Find the domain of the function $f(x)=\frac{\sqrt{2-x}}{x+5}$.

Solution: Start with the radical and then deal with the denominator. Set the contents of the radical to be greater than or equal to 0 and solve the inequality:

$2-x\geq 0$

$x\leq 2$

Now focus on the denominator of the function: $x + 5 \neq 0$, which means that $x \neq -5$. The domain of the function is the set of all real numbers less than or equal to 2, excluding –5.

We can write the domain of a function using interval notation. The table on page 14 summarizes the different types of intervals that we will encounter. Keep in mind that parentheses are always used next to the symbol for infinity, ∞. Brackets are never used next to ∞ because x can never actually reach infinity.

Interval	Name	Meaning	Examples
(a,b)	Open	All values of x that satisfy the inequality $a<x<b$ are contained in the interval. The endpoints are not contained in the interval.	$(1,2)$, $(-\infty,\infty)$, $(-\infty,2)$, $(3,\infty)$
$[a,b]$	Closed	All values of x that satisfy the inequality $a\le x\le b$ are contained in the interval. The endpoints are contained in the interval.	$[-2, 5]$
$(a,b]$	Half-open (or half-closed)	All values of x that satisfy the inequality $a<x\le b$ are contained in the interval. Only the right endpoint is contained in the interval.	$(-\infty,3]$, $(2,9]$
$[a,b)$	Half-open (or half-closed)	All values of x that satisfy the inequality $a\le x<b$ are contained in the interval. Only the left endpoint is contained in the interval.	$[-3,\infty)$, $[3,7)$

The domain of the function in Example 1, $f(x)=\frac{\sqrt{2-x}}{x+5}$, was the set of all real numbers less than or equal to 2, excluding –5. Using interval notation to describe this subset of the real numbers, we can write the domain as $(-\infty,-5)\cup(-5,2]$.

Lesson 1-2 Review

1. Write the domain of $f(x)=\frac{\sqrt{x+5}}{x-2}$ in interval notation.

Lesson 1-3: Operations on Functions

The *algebra* of real numbers establishes rules for how to combine real numbers. Numbers can be added, subtracted, multiplied, and divided. Algebraic expressions are an abstract way to represent numbers, so it is only natural that we are able to add, subtract, multiply, and divide algebraic expressions as well. There is one additional thing that we can do with functions: We can take their composition.

Think of a function as a transformation of things from the domain, or the input, to things in the range, or the output. If a function *f* has a domain X and its range is a subset of a set Y, then we use the notation $f\colon X \to Y$ to represent the idea that f is a function from X to Y. Suppose that g is a function whose domain is Y and whose range is a subset of a set Z. We can use the functions f and g to define a new function whose domain is X and whose range is contained in Z. This new function would take an element in X to its corresponding element in Y using the function f and then take that element in Y to an element in Z using the function g. This process of stringing functions from set to set is called the **composition** of functions. Because f takes things from X to Y, and g takes things from Y to Z, then the new function "g composed with f" takes things *directly* from X to Z.

We write the composition of f and g in the order described above as $g \circ f$. The functions are applied *right* to *left*: $g \circ f(x)$ means *first* apply the function f to x, and *then* apply the function g to the result. We can write $g \circ f(x) = g(f(x))$; $g \circ f(x)$ is read "g composed with f," or as "g of $f(x)$." The order in which we compose things matters. In general, $g \circ f(x) \neq f \circ g(x)$. In other words, $g(f(x)) \neq f(g(x))$.

We can look at a complicated function such as $h(x) = \sqrt{3x+1}$ in terms of the composition of two functions. If we define $f(x) = 3x + 1$ and $g(x) = \sqrt{x}$, then $h = g \circ f$. Functions are just instructions for what to do with the argument, or the object in parentheses. The function $g(x) = \sqrt{x}$ instructs us to take the argument of the function and put it under a radical. The function $f(x) = 3x + 1$ instructs us to triple the argument and then add 1. So:

$$h(x) = g(f(x)) = \sqrt{f(x)} = \sqrt{3x+1}$$

Alternatively, we could substitute in for $f(x)$ using its formula and then apply g:

$$h(x) = g(f(x)) = g(3x+1) = \sqrt{3x+1}$$

Either way you evaluate $g \circ f(x)$, you get the function $h(x)$.

Now let's look at the same composition in reverse order. Notice that, in this situation:

$$f \circ g(x) = f(g(x)) = 3g(x) + 1 = 3\sqrt{x} + 1$$

Alternatively, if we first substitute in for $g(x)$ using its formula and then apply f, we have:

$$f \circ g(x) = f(g(x)) = f(\sqrt{x}) = 3\sqrt{x} + 1$$

This illustrates the fact that the order in which you compose functions matters. In the example we just used, $g \circ f(x) = \sqrt{3x+1}$, while $f \circ g(x) = 3\sqrt{x} + 1$.

In general, $f \circ g(x)$ is a different function than $g \circ f(x)$.

Lesson 1-3 Review

Find $g \circ f(x)$ and $f \circ g(x)$ for the following pairs of functions:

1. $f(x) = \sqrt{6-x}$, $g(x) = 2x + 1$
2. $f(x) = \frac{1}{3x+6}$, $g(x) = \frac{1}{3}(x-6)$
3. $f(x) = x^2 + 2$, $g(x) = \sqrt{x-2}$

Lesson 1-4: Transformations of Functions

The graph of a function can be moved around the coordinate plane. The process by which a graph is moved is referred to as a **transformation** of a function. In general, the transformation of the graph of a function can involve a **shift** (also called a translation), a **reflection**, a **stretch**, or a **contraction.** Transformations can occur vertically, meaning that the result is a change in the value of the y-coordinate, or horizontally, meaning that they result in a change in the x-coordinate of the graph. We will look at each of these transformations individually.

To shift a function *vertically*, add a constant to the dependent variable, or to the function: The graph of $f(x) + c$ is the graph of $f(x)$ shifted up c units (if $c > 0$), or down c units (if $c < 0$). To shift a function *horizontally*, add a constant to the *argument* of the function: The graph of $f(x + c)$ is the graph of $f(x)$ shifted to the left c units (if $c > 0$), or to the right c units (if $c < 0$).

Think of reflecting a graph as looking at it in a mirror. Every point on one side of the mirror is sent to the corresponding point on the opposite side of the mirror. To reflect the graph of a function across the x-axis, multiply the function by -1: The graph of $-f(x)$ is the graph of $f(x)$ reflected across the x-axis. To reflect the graph of a function across the y-axis, the signs of the x-coordinates of every point of the function must be changed. In order to do this, the sign of the *argument* of the function must change. The graph of the function $f(-x)$ is the graph of $f(x)$ reflected across the y-axis.

The effect of stretching a graph is to "draw it out" so that it has the same shape but occupies more space. Contracting a graph effectively shrinks the graph, or makes it narrower. We can stretch or contract a graph vertically or horizontally. To *stretch* the graph of a function *vertically*, multiply the function by a constant that is greater than 1. To *contract* a function vertically, multiply the function by a positive constant that is less than 1 (for example, 1/2). To stretch the graph of a function horizontally, multiply the *argument* of the function by a positive constant that is less than 1. To contract the graph of a function, multiply the argument of the function by a constant that is greater than 1.

Lesson 1-4 Review

Transform the following functions:

1. Shift $f(x)=x^3$ to the right 4 units.
2. Shift $f(x)=3x-1$ up 8 units.
3. Shift $f(x)=\sqrt{x}$ to the left 2 units.
4. Shift $f(x)=\frac{1}{x}$ down 5 units.
5. Stretch $f(x)=\sqrt{x}$ vertically by a factor of 3.
6. Contract $f(x)=x^3$ horizontally by a factor of $\frac{1}{4}$

Lesson 1-5: Symmetry

The graph of a function can have some inherent symmetry. There are two important symmetries that we will discuss in this lesson. The first type of symmetry is when the y-axis serves as a mirror. The second type of symmetry has to do with the arrangement of the points of a function relative to the origin (the point at which x and y both equal zero).

In the graph of the function shown in Figure 1.1, the y-axis acts as a mirror. Every point to the left of the y-axis has a corresponding point to the right.

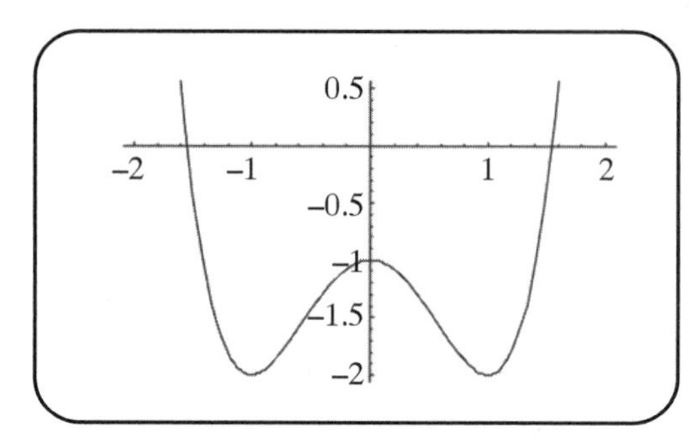

Figure 1.1.

The x-coordinates of a point on the left and its corresponding point on the right have the same magnitude; only their signs are different. The y-coordinates of a point on the left and its corresponding point on the right are exactly the same. Reflecting the graph of this function across the y-axis will not change the function. Recall that the transformation of reflecting across the y-axis can be written as $f(-x)$. Symmetry with respect to reflection across the y-axis can be characterized algebraically as: $f(-x) = f(x)$. If a function is symmetric with respect to the y-axis, that function is called an **even** function.

The graph of the function shown in Figure 1.2 is symmetric with respect to the origin. Notice that points in Quadrant I have corresponding points in Quadrant III and points in Quadrant II have corresponding points in Quadrant IV. This can be seen as reflecting a point across *both* the x-axis and the y-axis.

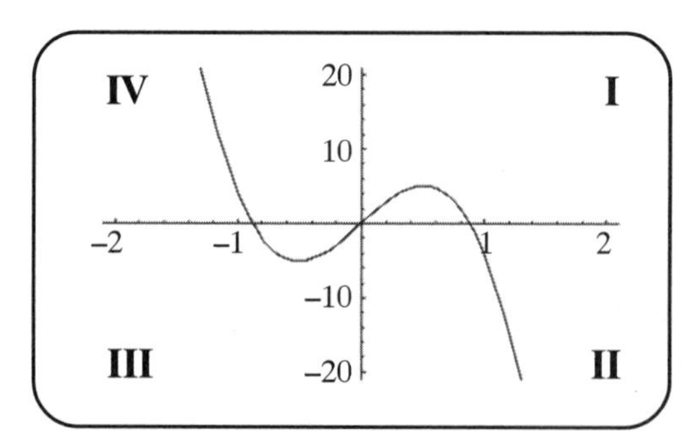

Figure 1.2.

Reflecting across the y-axis can be written as $f(-x)$, and reflecting across the x-axis can be written as $-f(x)$. Combining these transformations results in the overall transformation $-f(-x)$. The only point that doesn't change when you reflect it across both axes is the origin, which is why we refer to this type of symmetry as being symmetric with respect to the origin. A function that is symmetric with respect to the origin must satisfy the equation $f(x) = -f(-x)$. This is often rewritten as: $f(-x) = -f(x)$. Functions that are symmetric with respect to the origin are called **odd** functions.

Graphing a function that is symmetric requires less work if you use the symmetry to your advantage. You can plot half as many points and still see the whole picture. To graph an even function, find the y-intercept and then evaluate the function for some positive values of x. Using the symmetry of the function you can then graph the corresponding points to the left of the y-axis.

To graph an odd function, first realize that every odd function must pass through the origin. This can be seen by evaluating the equation $f(-x) = -f(x)$ when $x = 0$:

$$f(-x) = -f(x)$$

$$f(-0) = -f(0)$$

$$f(0) = -f(0)$$

The only number that is unchanged when you multiply by -1 is 0, so $f(0) = 0$ and we see that an odd function must pass through the origin. To graph an odd function, evaluate the function for some positive values of x. Using the symmetry of the function and the fact that the graph of the function must pass through the origin, you can graph the corresponding points.

Lesson 1-6: Difference Quotients

The difference quotient is very important in calculus. It forms the basis of the definition of the derivative, and the derivative will appear throughout this book. The difference quotient is an abstract way of talking about slopes and average rates of change. Evaluating the difference quotient involves evaluating functions when the argument is replaced by an algebraic expression.

In Chapter 7 we will explore the geometric significance of the difference quotient. For now, we will focus on the algebraic aspects of the difference quotient and practice simplifying it. Given any function $f(x)$, the **difference quotient** of the function at a point x is defined to be the ratio $\frac{f(x+h)-f(x)}{h}$. The difference quotient of a function $f(x)$ is actually a function of two variables: x and h. Don't let the two variables bother you. Each variable represents a number, and, as we will see later on, we will usually want h to be a number that is small in magnitude. Looking at the difference quotient expression $\frac{f(x+h)-f(x)}{h}$, we can see right away that there will be problems if h is ever equal to 0. In fact, one of our requirements is that $h \neq 0$. One of the games we will play as we learn calculus is to let h be as small as possible, without actually becoming 0. It's kind of like letting h dance the limbo and asking, "How low can it go?" We usually approach difference quotients in three steps:

Step 1. Evaluate $f(x + h)$.

Step 2. Evaluate $f(x + h) - f(x)$.

Step 3. Put the pieces together and evaluate $\frac{f(x+h)-f(x)}{h}$, simplifying the expression as much as possible.

Example 1

Let $f(x) = 3x + 4$. Find the difference quotient $\frac{f(x+h)-f(x)}{h}$.

Solution: Step 1: Evaluate this function when the argument is $x + h$:

$$f(x) = 3x + 4$$

$$f(x+h) = 3(x+h) + 4 = 3x + 3h + 4$$

Step 2: Evaluate $f(x+h) - f(x)$:

$$\begin{aligned} f(x+h) - f(x) &= (3x + 3h + 4) - (3x + 4) \\ &= 3x + 3h + 4 - 3x - 4 \\ &= 3h \end{aligned}$$

Step 3: Evaluate and simplify the difference quotient. In Step 2 we simplified the numerator, and we can substitute the simplified form of $f(x+h) - f(x)$ directly into the numerator of $\frac{f(x+h)-f(x)}{h}$:

$$\frac{f(x+h)-f(x)}{h}=\frac{3\cancel{h}}{\cancel{h}}=3 \text{, as long as } h \neq 0.$$

Let me emphasize that we are allowed to cancel the h's in the difference quotient *as long as we stipulate that h ≠ 0*!

Example 2

Let $f(x)=x^2+3x$. Evaluate the difference quotient

$$\frac{f(x+h)-f(x)}{h}.$$

Solution: Step 1: Evaluate this function when the argument is $x+h$:

$$f(x)=x^2+3x$$

$$\begin{aligned} f(x+h) &= (x+h)^2+3(x+h) \\ &= x^2+2xh+h^2+3x+3h \end{aligned}$$

Step 2: Evaluate $f(x+h)-f(x)$:

$$\begin{aligned} f(x+h)-f(x) &= \left(x^2+2xh+h^2+3x+3h\right)-\left(x^2+3x\right) \\ &= x^2+2xh+h^2+3x+3h-x^2-3x \\ &= 2xh+h^2+3h \end{aligned}$$

Step 3: Evaluate and simplify the difference quotient. In Step 2 we simplified the numerator, and we can substitute the simplified form of $f(x+h)-f(x)$ directly into the numerator of $\frac{f(x+h)-f(x)}{h}$:

$$\begin{aligned} \frac{f(x+h)-f(x)}{h} &= \frac{2xh+h^2+3h}{h}=\frac{\cancel{h}(2x+h+3)}{\cancel{h}} \\ &= 2x+h+3 \end{aligned}$$

as long as $h \neq 0$.

We will revisit the difference quotient when I introduce the concept of the derivative. Until then, practice working with the difference quotient and becoming familiar with the algebraic steps involved in simplifying it. The time you spend practicing now will pay off down the road.

Lesson 1-6 Review

Evaluate the difference quotient for the following functions:

1. $f(x) = 6x - 5$
2. $f(x) = 2x^2 - x$

Lesson 1-7: Increasing and Decreasing Functions

When analyzing a function, it is important to know where the graph of the function is rising and where it is falling. A function is increasing if the value of the function increases as the value of the independent variable increases. We can state this definition more precisely:

> A function $f(t)$ is **increasing** on an interval I if $f(a) < f(b)$ whenever $a < b$.

Similarly, a function is decreasing if the value of the function decreases as the value of the independent variable increases. This can also be stated precisely:

> A function $f(t)$ is **decreasing** on an interval I if $f(a) > f(b)$ whenever $a < b$.

A function is **monotonic** on an interval if it is either always increasing or always decreasing on that interval. In other words, a monotonic function does not change directions. Linear functions are monotonic on their entire domain. A linear function will be monotone increasing if it has a positive slope, and it will be monotone decreasing if it has a negative slope.

Fortunately, the definitions for increasing and decreasing functions do not involve advanced mathematics. A comparison of function values at various points in the domain will determine whether or not a function is increasing, decreasing, or neither, on a particular interval I. Unfortunately, evaluating a function at a variety of points in the domain can become tedious. We can, however, use calculus to more efficiently determine the regions where a function is increasing and the regions where it is decreasing.

Answer Key

Lesson 1-2 Review

1. The domain of $f(x)=\frac{\sqrt{x+5}}{x-2}$ is $x\geq -5$ and $x\neq 2$.

   This can be written in interval notation as $[-5,2)\cup(2,\infty)$.

Lesson 1-3 Review

1. $g\circ f(x)=g(f(x))=2f(x)+1=2\sqrt{6-1}+1$

   $f\circ g(x)=f(g(x))=\sqrt{6-g(x)}=\sqrt{6-(2x+1)}=\sqrt{5-2x}$

2. $g\circ f(x)=g(f(x))=\frac{1}{3}\left(\frac{1}{3x+6}-6\right)=\frac{1}{3}\left(\frac{1}{3x+6}-\frac{6(3x+6)}{(3x+6)}\right)=\frac{1}{3}\left(\frac{-18x-35}{3x+6}\right)$

   $f\circ g(x)=f(g(x))=\frac{1}{3g(x)+6}=\frac{1}{3\left(\frac{1}{3}(x-6)\right)+6}=\frac{1}{(x-6)+6}=\frac{1}{x}$

3. $g\circ f(x)=g(f(x))=\sqrt{f(x)-2}=\sqrt{(x^2+2)-2}=\sqrt{x^2}=x$

   $f\circ g(x)=f(g(x))=[g(x)]^2+2=\left(\sqrt{x-2}\right)^2+2=(x-2)+2=x$

Lesson 1-4 Review

1. $f(x-4)=(x-4)^3$

2. $f(x)+8=(3x-1)+8=3x+7$

3. $f(x+2)=\sqrt{x+2}$

4. $f(x)-5=\frac{1}{x}-5$

5. $3f(x)=3\sqrt{x}$

6. $f(4x)=(4x)^3$

Lesson 1-6 Review

1. $f(x)=6x-5$, $f(x+h)=6(x+h)-5$,

   $\frac{f(x+h)-f(x)}{h}=\frac{(6(x+h)-5)-(6x-5)}{h}=\frac{6x+6h-5-6x+5}{h}=\frac{6\cancel{h}}{\cancel{h}}=6$ if $h\neq 0$

2. $f(x) = 2x^2 - x$,

$$f(x) = 2(x+h)^2 - (x+h) = 2(x^2 + 2xh + h^2) - x - h$$
$$= 2x^2 + 4xh + 2h^2 - x - h$$

$$\frac{f(x+h) - f(x)}{h} = \frac{(2x^2+4xh+2h^2-x-h)-(2x^2-x)}{h} = \frac{4xh+2h^2-h}{h} = \frac{\not{h}(4x+h-1)}{\not{h}}$$
$$= 4x + h - 1 \text{ if } h \neq 0$$

Elementary Functions

During the course of our study of calculus, we will be analyzing a variety of functions. There are a few elementary functions that we should understand in detail. In this chapter, we will identify some of these functions and *briefly* review their properties. Familiarity with these elementary functions will make your calculus explorations much more meaningful.

Lesson 2-1: Linear Functions

The first group of elementary functions that we will examine are linear functions. **Linear functions** are functions of the form $f(x) = mx + b$, where m is called the **slope** and b is the ***y*-intercept** of the function. An example of a linear function is $f(x) = 2x + 3$. For linear functions, $m \neq 0$. For the special case where $m = 0$, the function $f(x) = b$ is called a **constant function.** A constant function is *not* a linear function. The graph of a *linear* function is a line that is neither vertical nor horizontal. The graph of a *constant* function is a horizontal line, and a vertical line is not a function at all. The domain of a linear function is the set of all real numbers. A linear function is completely determined by two pieces of information: its slope and its y-intercept.

In order to find the equation of a line, we need to know two things: a point that the line passes through, and its slope. From that information we can use the point-slope formula to write the equation of a line in slope-intercept form. The equation of a line that passes through the point (x_1, y_1) with slope m is given by: $y - y_1 = m(x - x_1)$

It can be simplified and written in the form $y = mx + b$.

Replacing y with $f(x)$ yields the familiar linear function $f(x) = mx + b$. As I mentioned in Chapter 1, the intercepts of a function are very important points. All *linear* functions have exactly one x-intercept and one y-intercept.

Linear functions are useful in mathematical modeling. They are fairly easy to evaluate and are uniquely determined using only two points. They can be used to extrapolate outside of a data set, and they can be used to interpolate between points in a data set.

Example 1

A minor league baseball team plays in a park with a seating capacity of 10,000 spectators. With the ticket price set at \$10, the average attendance at recent games has been 7,000 spectators. A market survey indicates that for each dollar the ticket price is lowered, the average attendance increases by 800 spectators. Find a linear function that models the attendance as a function of price, and find the price that will lead to a full house.

Solution: We need to express the attendance as a linear function of price. Attendance represents the dependent variable, and price represents the independent variable. We need to use variables for attendance and price. Let a represent attendance and p represent price. We are looking for the function $a(p)$. To find the equation of a line, we need two points. We are given one data point, corresponding to the point (10, 7000), and instructions on how to find a second point. If the price is lowered by \$1 (meaning that the price is now \$9), the attendance will be 7,800. The point (9, 7800) is another data point. With two points, we can find the equation of the line that passes through them, but we first need to find the slope:

$$\text{slope} = \frac{\Delta a}{\Delta p} = \frac{7{,}800 - 7{,}000}{9 - 10} = -800$$

The fact that the slope is –800 should not be a surprise: A *decrease* in price of \$1 results in an *increase* in attendance of 800. Now that we know the slope and a point that the line passes through, we can find the equation of the line. We can use either point to find the equation of a line, and I will use the point (10, 7000):

$$a - 7{,}000 = -800(p - 10)$$

$$a = -800p + 15{,}000$$

This equation represents the linear function that models attendance as a function of price. If the house is full, then there will be 10,000 spectators in attendance. We can use our model to solve for the price:

$$\begin{aligned} a &= -800p + 15{,}000 \\ 10{,}000 &= -800p + 15{,}000 \\ -5{,}000 &= -800p \\ p &= 6.25 \end{aligned}$$

Charging a price of \$6.25 will fill the house, according to our model.

It is important to realize the limitations of our mathematical models. Many variables besides price will affect attendance at a baseball game, but introducing these other variables will result in a much more complicated model. Linear models are a good place to start, despite their simplistic nature.

Lesson 2-1 Review

1. Find the linear function that passes through the points (2, 5) and (5, 7).
2. Find the intercepts of the linear function $f(x) = 3x - 4$.
3. The cost for printing T-shirts for a local non-profit organization includes \$50 for the T-shirt design and \$4 per shirt. Express the cost of printing the T-shirts as a linear function of the number of T-shirts that are printed, and determine the cost of printing 500 T-shirts.

Lesson 2-2: Quadratic Functions

A quadratic function is a function of the form $f(x) = ax^2 + bx + c$, where a, b, and c are constants, and $a \neq 0$. The function $f(x) = 3x^2 + 2x + 1$ is one example of a quadratic function. The constant a is called the **leading coefficient** of the quadratic function, and b is called the **linear coefficient.** The graph of a quadratic function is called a **parabola.** The domain of a quadratic function is the set of all real numbers. There are five important features of a quadratic function, and analyzing these features will help us to easily graph these types of functions.

The first feature of a quadratic function $f(x) = ax^2 + bx + c$ is that its graph will either "open" up or down, depending on the sign of the leading coefficient. If $a > 0$ the parabola will open up, or look as if it bends upward, and if $a < 0$ the parabola will open down, or look as though it bends downward. We refer to the direction of a curve as its **concavity.** A parabola that opens upward is called **concave up**, and a parabola that opens downward is called **concave down**. The graph of the quadratic function $f(x) = 2x^2 - x + 1$ is a parabola that opens upward, or is concave up, and the graph of the quadratic function $f(x) = -3x^2 + x - 7$ is a parabola that opens downward, or is concave down. Figure 2.1 shows the graphs of two parabolas: one that opens up and another that opens down.

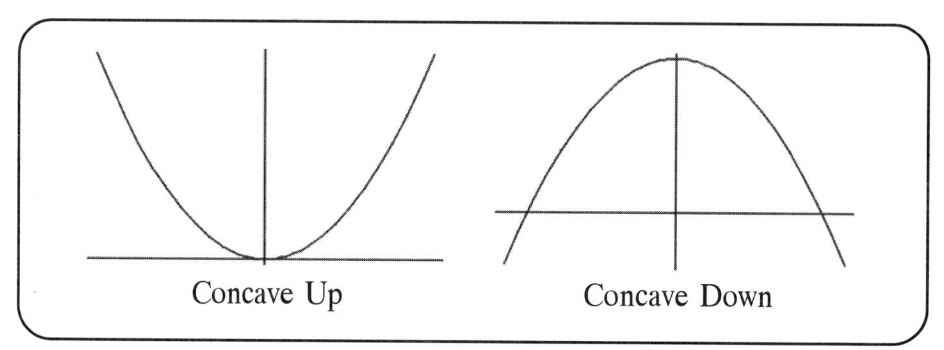

Figure 2.1.

The second feature of a quadratic function $f(x) = ax^2 + bx + c$ has to do with the symmetry of its graph. Every quadratic function has a vertical line that splits it into two symmetrical halves. This vertical line is called the **axis of symmetry**. The equation for the axis of symmetry is:

$$x = -\frac{b}{2a}$$

The third feature of a quadratic function has to do with the point where the axis of symmetry intersects the graph of the parabola. This point is called the **vertex** of the parabola. If the parabola opens up, then the vertex corresponds to the *lowest* point on the parabola, or the **absolute minimum** value that the function achieves on its domain. If the parabola opens down, then the vertex corresponds to the *highest* point on the parabola, or the **absolute maximum** value that the function achieves on its domain. The vertex of a parabola is the point:

$$\left(-\tfrac{b}{2a}, f\left(-\tfrac{b}{2a}\right)\right).$$

The fourth feature of a quadratic function is the location of the y-intercept. Every quadratic function has a y-intercept, and it is located at the point $(0, f(0))$. The y-intercept of the quadratic function $f(x) = ax^2 + bx + c$ is the point $(0, c)$.

The fifth, and final, feature of a quadratic function has to do with the existence and the location of the x-intercepts of the function. The x-intercepts of a function are where the function intersects the x-axis. These points are found by solving the equation $f(x) = ax^2 + bx + c = 0$. The three main methods for solving quadratic equations include factoring, completing the square, and using the quadratic formula.

The quadratic formula gives the roots, or zeros, of the quadratic function $f(x) = ax^2 + bx + c$. Equivalently, the quadratic formula specifies the location of the x-intercepts of the function $f(x) = ax^2 + bx + c$. "Roots," "zeros," and "x-intercepts" are all terms that describe the same thing. Written yet another way, the quadratic formula can be used to write the solutions to the equation $ax^2 + bx + c = 0$:

$$x = \frac{-b \pm \sqrt{b^2 - 4ac}}{2a}$$

You can solve *any* solvable quadratic equation by using the quadratic formula. The key is to be able to recognize whether or not a quadratic equation is solvable. Not all parabolas have any x-intercepts. A parabola will have x-intercepts if the corresponding quadratic equation has a solution. One indication of whether a quadratic equation is solvable is to evaluate the discriminant. Given the quadratic equation $ax^2 + bx + c = 0$, the **discriminant** is the value:

$$b^2 - 4ac$$

The quadratic equation will be solvable as long as the discriminant is *not* a negative number.

Quadratic functions can be used to model a variety of situations, including the trajectory of a ball thrown in the air and the revenue generated from the sale of a product.

Example 1

A minor league baseball team plays in a park with a seating capacity of 10,000 spectators. With the ticket price set at $10, the average attendance at recent games has been 7,000. A market survey indicates that for each dollar the ticket price is lowered, the average attendance increases by 800 spectators. Using a linear function to model the attendance as a function of price, find a quadratic model for the *revenue* generated as a function of the price of a ticket. Use that model to find the ticket price that will maximize the revenue.

Solution: In Example 1 of Lesson 2-1, we found a linear model for the attendance as a function of the price:

$$a = -800p + 15{,}000$$

We can use that model to create a function for the revenue generated from our ticket sales as a function of the price of the ticket. Let R represent the revenue, and p represent price. The revenue generated from ticket sales is found by taking the product of the number of tickets sold times the price per ticket. The number of tickets sold is the attendance, and the price per ticket is p. We can use our equation for attendance as a function of price to simplify this revenue equation:

$$\begin{aligned} R &= a \cdot p \\ &= (-800p + 15{,}000) \cdot p \\ &= -800p^2 + 15{,}000p \end{aligned}$$

Our equation for revenue is a parabola that is concave down, because the leading coefficient is negative. We can find the maximum possible revenue by finding the p-coordinate of the vertex of the parabola:

$$p = \frac{-(15{,}000)}{2(-800)} = 9.375$$

We can round our answer to $9.38.

In the previous lesson, we saw that the ticket price to "fill the house" was $6.25, but that price did not generate the most revenue.

If the graph of a parabola is concave up and has a vertex located at $(v, f(v))$, then the quadratic function will be *decreasing* on the interval $(-\infty, v)$ and it will be *increasing* on the interval (v, ∞). If the graph of a parabola is concave down and has a vertex located at $(v, f(v))$, then the quadratic function will be *increasing* on the interval $(-\infty, v)$ and it will be *decreasing* on the interval (v, ∞).

Example 2

Determine the interval(s) on which $f(x) = -2x^2 + 3x - 5$ is increasing.

Solution: The graph of this quadratic function is concave down, so it will be increasing on the interval $(-\infty, v)$, where v is the x-coordinate of the vertex. The x-coordinate of the vertex has the same value as the axis of symmetry:

$$x = \frac{-b}{2a} = \frac{-3}{2(-2)} = \frac{3}{4}$$

The function $f(x) = -2x^2 + 3x - 5$ is increasing on the interval $\left(-\infty, \frac{3}{4}\right)$.

In addition to graphing a quadratic function, it is useful to be able to solve quadratic inequalities. There are four types of quadratic inequalities:

$$ax^2 + bx + c > 0 \qquad ax^2 + bx + c < 0$$

$$ax^2 + bx + c \geq 0 \qquad ax^2 + bx + c \leq 0$$

One procedure for solving a quadratic inequality is outlined here. We will build on this procedure when we discuss how to graph polynomials and rational functions.

1. Rearrange the terms in the quadratic inequality so that there is a 0 on one side of the inequality, and all other terms are on the other side.
2. Factor the quadratic expression and set each factor equal to zero find the zeros (or roots) of the quadratic expression. The zeros of a quadratic function are the only places where the function *can* change sign. If there are no real zeros (because the discriminant of the quadratic expression is negative), then the quadratic function will never change sign.
3. Arrange the zeros on a number line, effectively dividing the number line into regions separated by the zeros of the quadratic expression.
4. Create a sign chart for the function. For each region, pick a "test value" or a number in the region and evaluate the quadratic expression at each of the test values. The sign of the quadratic expression at a test value indicates the sign of the quadratic function throughout the entire region.

5. Pick out the regions that satisfy the quadratic inequality. If the inequality is strict ($<$ or $>$), do not include the endpoints of the regions. If the inequality is not strict ($\leq$ or $\geq$), include the endpoints of the regions.

Example 3

Solve the quadratic inequality $x^2 \leq x + 6$.

Solution: Follow the procedure outlined above:

- Rearrange the inequality:
  $x^2 \leq x + 6$ is equivalent to $x^2 - x - 6 \leq 0$:
- Factor the quadratic expression: $(x - 3)(x + 2)$.
  The zeros are $x = 3$ and $x = -2$.
- Draw a number line and label the points $x = 3$ and $x = -2$.
- Create a sign chart. The two roots of the quadratic expression will divide the number line into three regions. Choose a test value from each of the regions. The test values I will use are $x = -3$, $x = 0$, and $x = 4$. (The test values you choose may be different than the ones I choose. It doesn't matter.) The goal is to determine the sign of the quadratic expression at these test values, and if everything is done carefully, your sign chart will agree with my sign chart: $f(-3) = 6 > 0$, $f(0) = -6 < 0$ and $f(4) = 6 > 0$. The signs of the quadratic expression are shown in Figure 2.2.

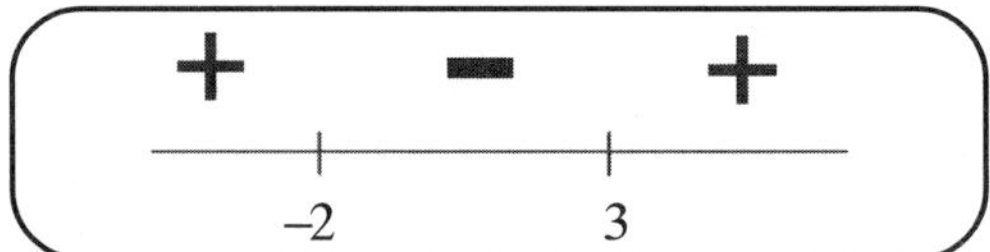

Figure 2.2.

- The quadratic inequality we are trying to solve is $x^2 - x - 6 \leq 0$. This is not a strict inequality, so we will include the endpoints of the interval. From Figure 2.2, we see that the region that satisfies the inequality is the interval $[-2, 3]$.

Another way to solve the inequality $x^2 \leq x + 6$ involves graphing the parabola $f(x) = x^2$ and the line $g(x) = x + 6$ on the same axes, as shown in

Figure 2.3. Find the points of intersection of the two graphs, and locate the region where the graph of the parabola lies below the graph of the line. The values of x that lie in that region are the solutions to the inequality. This graphical technique can be used to solve inequalities that involve a mixture of functions, such as exponential, logarithmic, polynomial, or trigonometric functions.

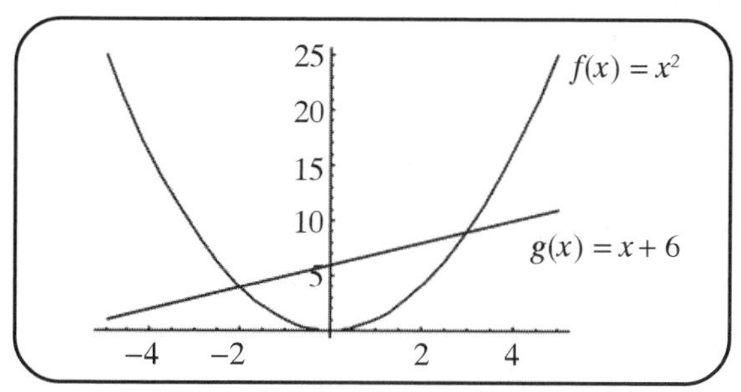

Figure 2.3.

Lesson 2-2 Review

Address the five features of the following quadratic functions:

1. $f(x)=x^2-8x+12$

2. $f(x)=-x^2+6x-4$

3. Find the intervals where the following functions are increasing, and the intervals where they are decreasing:

   a. $f(x)=x^2+2x$

   b. $g(x)=-3x^2-2x+1$

4. Solve the quadratic inequality $x^2<2x+15$.

Lesson 2-3: Polynomials

Linear and quadratic functions are examples of a general class of functions called polynomials. A **polynomial** is a function of the form $f(x)=a_nx^n+a_{n-1}x^{n-1}+...+a_1x+a_0$. The coefficients $a_n, a_{n-1}, ..., a_1, a_0$ are real numbers with $a_n \neq 0$. The constant a_n is called the **leading coefficient**, and a_0 is referred to as the **constant coefficient**, or the **constant term**. The value of n must be a non-negative integer, and it is called the **degree** of the polynomial. The domain of a polynomial function is the set of all real numbers.

The graph of a polynomial is related to its degree. You already have some experience graphing polynomials, and in Chapters 13 and 14 we will build on that experience by introducing calculus techniques to analyze the graph of a function in more detail. Right now, there are a few quick observations to make about the graphs of polynomials in general.

Figure 2.4 shows the graphs of some polynomials of increasing degree. The first polynomial has degree 1, the second has degree 2, and so on. The degree of the sixth polynomial is 6. Notice that the polynomials with odd degree share some common characteristics, and the polynomials with even degree also have common properties. We will examine these graphs in more detail.

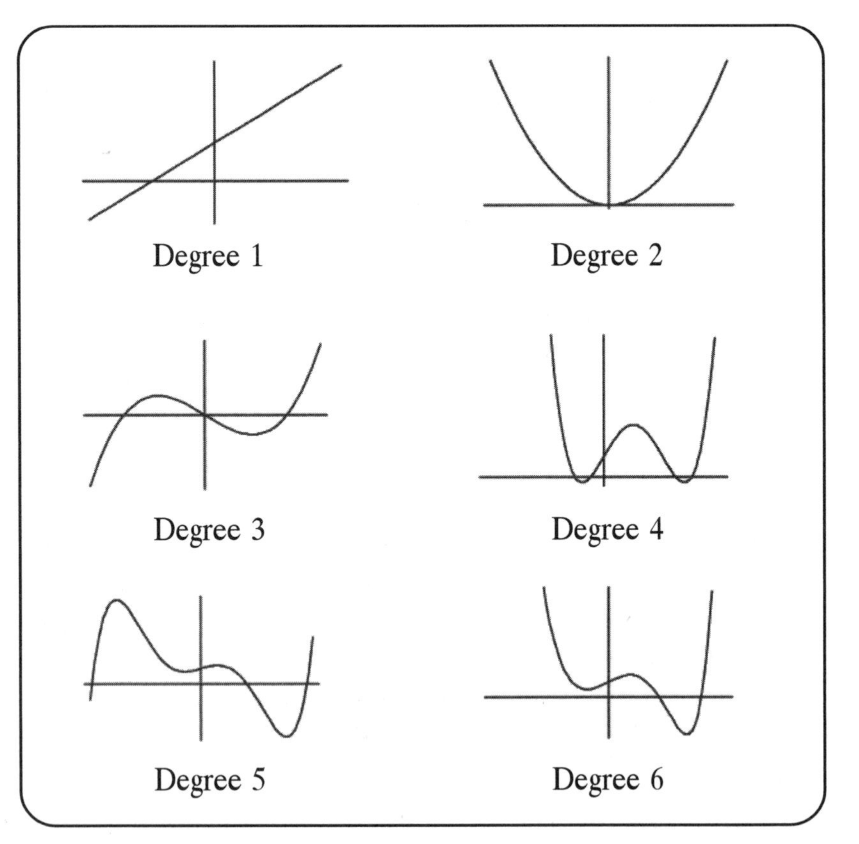

Figure 2.4.

With all of the polynomials, there are no breaks in their graphs. Polynomials are examples of functions that are **continuous**. We will learn more about continuous functions in Chapter 6. Polynomials are also **smooth**, in that there are no sharp corners or cusps.

Polynomials of odd degree all cross the x-axis at least once. If they start out negative, they will eventually end up positive, and vice versa. A polynomial of odd degree may or may not have a **turning point**, or a point where the function changes direction from increasing to decreasing, or vice versa.

Polynomials of even degree may or may not cross the x-axis, but they always end up with the same sign that they started with. If they start out negative, they may or may not cross the x-axis and become positive, but they eventually end up negative. If they start out positive, they may or may not cross the x-axis and become negative, but they eventually end up positive. Notice that each of the polynomials of even degree has at least one turning point.

In general, a polynomial of degree n will cross the x-axis at most n times, or have at most n x-intercepts, and will have at most $(n - 1)$ turning points. Finding the locations of the x-intercepts and finding the turning points actually involves the same procedure. The analytical techniques that we learn in calculus will provide a systematic way to locate these important points.

Polynomials can also exhibit certain symmetries. If a polynomial only involves odd powers of x and passes through the origin, so that $a_0 = 0$, then the graph of the polynomial will be symmetric about the origin. In other words, a polynomial that only involves *odd* powers and has a constant coefficient equal to 0 is an *odd* function. Similarly, if a polynomial only involves even powers of x, then the graph of the polynomial will be symmetric about the y-axis. In other words, a polynomial that only involves *even* powers is an *even* function. There is no special restriction on the constant coefficient for an even function: a_0 does not have to be 0, as even functions do not have to pass through the origin.

Combining even and odd powers of x destroys any of the even/odd symmetry just mentioned. A polynomial that involves both even and odd powers of x will not be even, because of the odd powers, and it will not be odd, because of the even powers. Also, there is only *one* polynomial that is both even and odd: the function $f(x) = 0$.

Lesson 2-3 Review

1. Determine whether the following polynomials are even, odd, or neither.

   a. $f(x) = 3x^4 - 2x + 1$

   b. $f(x) = 3x^{10} - 2x^6 + 5$

   c. $f(x) = 3x^9 - 2x^5 + 1$

   d. $f(x) = 3x^9 - 2x^5 + x$

Lesson 2-4: Rational Functions

A rational function is a function that is the ratio of two polynomials and can be written as $f(x) = \frac{P(x)}{Q(x)}$, where $P(x)$ and $Q(x)$ are polynomials. An example of a rational function is $f(x) = \frac{x+1}{x^2+1}$. One of the important features of a rational function is its **asymptotic behavior.** An asymptote is a line (it can be horizontal, vertical, or slanted) that a function approaches, or gets arbitrarily close to. A rational function can have vertical, horizontal, or oblique asymptotes. A rational function will have a **vertical asymptote** if the magnitude of the function gets arbitrarily large, or tends to infinity, as x approaches some fixed, finite number. A rational function will have a **horizontal asymptote** if the function approaches some fixed, finite number as the magnitude of x gets arbitrarily large, or tends to infinity. A rational function will have an **oblique asymptote** if the degree of the polynomial in the numerator is one more than the degree of the polynomial in the denominator. In order for a function to have a vertical or a horizontal asymptote, the magnitude of one of the variables, either the dependent or the independent variable, must head towards infinity while the other variable heads towards a finite number.

The vertical asymptotes of a rational function are fairly easy to recognize. Factor both the numerator and the denominator of the rational function, and find the zeros of the denominator (by setting each factor in the denominator equal to 0 and solving for x). The rational function will have a vertical asymptote at any value of x that is a zero of the denominator and *not* a zero of the numerator. For the function $f(x) = \frac{(x-2)(x+3)}{(x+1)(x-4)}$, the denominator is 0 when $x = -1$ or $x = 4$, and the zeros of the denominator are not zeros of the numerator. The function $f(x)=\frac{(x-2)(x+3)}{(x+1)(x-4)}$ has two vertical asymptotes: $x = -1$ and $x = 4$.

A rational function can have at most one horizontal asymptote. To find the horizontal asymptote of a rational function, we first need to understand the nature of a polynomial as the magnitude of the independent variable becomes large. As x becomes large, only the leading term of a polynomial becomes important, and will dominate the behavior of the polynomial.

To find the horizontal asymptote of a rational function, focus on the leading term of each of the two polynomials that make up the rational function. If $R(x)$ is a rational function, we can write $R(x)$ as a ratio of two polynomials:

$$R(x)=\frac{a_n x^n + a_{n-1}x^{n-1} + \ldots + a_1 x + a_0}{b_m x^m + b_{m-1}x^{m-1} + \ldots + b_1 x + b_0}$$

When x is large, $R(x) \sim \frac{a_n x^n}{b_m x^m}$. (I am using the symbol $\sim$ to mean "behaves like.") There are three cases that we need to consider.

1. The degree of the numerator is greater than the degree of the denominator, or $n > m$. In this case, we can cancel some of the powers of x and $R(x) \sim \frac{a_n}{b_m} x^{n-m}$, with $n - m > 0$. As x gets large, the magnitude of this rational function also gets large, so there is no horizontal asymptote.
2. The degree of the numerator equals the degree of the denominator, or $n = m$. In this case, $R(x) \sim \frac{a_n}{b_n}$, and the rational function gets close to the value of the ratio of the leading coefficients. The horizontal asymptote will be the line $y = \frac{a_n}{b_n}$.
3. The degree of the numerator is less than the degree of the denominator, or $n < m$. In this case, we can cancel out some

of the powers of x and $R(x) \sim \frac{a_n}{b_m x^{m-n}}$, with $m-n>0$. As x gets large, the value of $R(x)$ heads towards 0, and the horizontal asymptote will be the line $y=0$.

The intercepts of a rational function are also important points. The y-intercept of a rational function is found by evaluating the function at $x = 0$. Of course, if $x = 0$ is not in the domain of the rational function (because the function has a vertical asymptote at $x = 0$), then the rational function will not have a y-intercept. The x-intercepts of a rational function are the zeros of the polynomial in the numerator. If the numerator and denominator of a rational function have already been factored (to find the vertical asymptotes of the function), then most of the work necessary to find the x-intercepts has already been done. This will be illustrated in Example 1.

Example 1

Find the intercepts and the asymptotes of the following rational functions:

a. $R(x)=\frac{x^2-2x+1}{2x^2+2x-12}$ b. $F(x)=\frac{(x-2)(x+3)}{2(x+1)^2(x-1)}$

Solution:

a. The y-intercept is the point $(0, R(0))$:

$$R(0)=\frac{0^2-2\cdot 0+1}{2\cdot 0^2+2\cdot 0-12}=-\frac{1}{12}$$

The y-intercept is the point $\left(0,-\frac{1}{12}\right)$

The horizontal asymptote: The degree of the numerator and the degree of the denominator are the same, so

$$R(x)\sim\frac{x^2}{2x^2}=\frac{1}{2}$$

Therefore, the horizontal asymptote is the line

$$y=\frac{1}{2}$$

To find the x-intercepts and the vertical asymptotes, we will need to factor the numerator and denominator of $R(x)$:

$$R(x)=\frac{(x-1)^2}{2(x+3)(x-2)}$$

The x-intercept: Find the zeros of the numerator.
There is only one x-intercept: (1, 0)

The vertical asymptotes: Locate the zeros of the denominator and check to see if they are also zeros of the numerator. The zeros of the denominator are $x = -3$ and $x = 2$, neither of which are zeros of the numerator.

Therefore the vertical asymptotes of the function $R(x)$ are

$$x = -3 \text{ and } x = 2$$

b. The y-intercept is the point $(0, F(0))$:

$$F(0)=\frac{(0-2)(0+3)}{2(0+1)^2(0-1)}=\frac{-6}{-2}=3$$

The y-intercept is the point (0, 3)

The horizontal asymptote: The degree of the numerator is 2 and the degree of the denominator is 3, so the horizontal asymptote is the line $y = 0$

The polynomials in the numerator and the denominator are already factored, so most of the work in finding the x-intercepts and the asymptotes has already been done.

The x-intercepts are the points (2, 0) and (–3, 0)

The vertical asymptotes: Locate the zeros of the denominator that are not zeros of the numerator. The zeros of the denominator of $F(x)$ are $x = -1$ and $x = 1$, neither of which are zeros of the numerator.

Therefore the vertical asymptotes of $F(x)$ are

$$x = -1 \text{ and } x = 1$$

Lesson 2-4 Review

Find the intercepts and the asymptotes of the following functions:

1. $G(x)=\dfrac{x^2+x-6}{2x^2-x-3}$

2. $H(x)=\dfrac{x^2-1}{x^3-5x^2+6x}$

Lesson 2-5: Power Functions

A **power function** is a function of the form $f(x) = ax^p$, where a and p are real numbers. An example of a power function is $y = 3x^{1/2}$. The domain of a power function depends on the constant p. A power function with $p = 0$ is a constant function: $f(x) = a$. The graph of this function is a horizontal line. In this lesson, we will examine power functions for non-zero values of p.

If $p \geq 0$, then $x = 0$ will be in the domain of the function. If $p < 0$, then $x = 0$ will not be in the domain of the function. Strange things happen if p is not an integer. Problems will occur if you take an even root of a negative number, or if you raise a negative number to an irrational power. For the analysis of power functions in this section, we will restrict our domain to the non-negative real numbers. We can use the appropriate symmetry of a power function to analyze the function over the negative real numbers, if the negative real numbers are included in the domain of the power function.

We will first consider power functions of the form $f(x) = x^p$ for positive values of p that are odd. Figure 2.5 shows the graphs of the functions $f(x) = x$, $f(x) = x^2$, and $f(x) = x^3$ on the same axes, for values of x between 0 and 2.

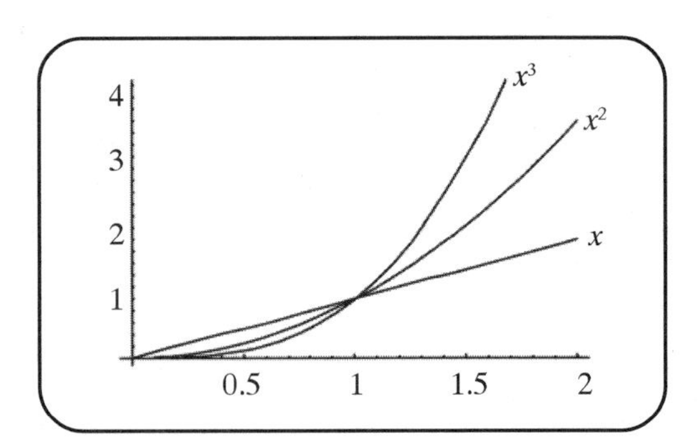

Figure 2.5.

These functions all pass through the origin and the point (1, 1). Notice that on the interval (0, 1), larger values of p result in graphs that are lower. For values of x greater than 1, the opposite trend holds: Higher values of p result in graphs that are higher.

Power functions with an even integer power (x^2, x^4, x^6, and so on) are always concave up, and they have an absolute minimum at $x = 0$. Power functions with an odd integer power (x^3, x^5, and so on) do not have an absolute minimum or an absolute maximum, but they do change concavity. Power functions with an odd integer power are concave down for negative values of x and concave up for positive values of x.

Power functions with negative integer powers (x^{-1}, x^{-2}, and so on) can be analyzed using our techniques for analyzing rational functions. They all have a vertical asymptote at $x = 0$, a horizontal asymptote $y = 0$, and no intercepts. If the magnitude of the power is even, the graph of the function will be concave up on its domain, and the function will be symmetric about the y-axis. If the magnitude of the power is odd, the graph of the function will be concave down if $x < 0$ and concave up if $x > 0$, and the function will be symmetric about the origin.

An interesting group of power functions to study have powers that are positive rational numbers. The function $f(x) = x^p$ is called the pth **power** of x, and $g(x) = x^{1/p}$ is called the pth **root** of x. Some fractional powers involve roots that are only defined for non-negative values of x, so we will restrict the domain of these functions to $x \geq 0$. Figure 2.6 shows the graphs of the functions $f(x) = x^2$, $f(x) = x^{3/2}$, $f(x) = x$, $f(x) = x^{1/2}$, and $f(x) = x^{1/3}$. All of these functions pass through the points (0, 0) and (1, 1). For values of x that are greater than 1, the higher powers result in higher graphs. For values of x that are less than 1, the opposite trend holds: The smaller powers result in higher graphs. Notice also that the power functions with $p > 1$ are concave up, and the power functions with $0 < p < 1$ are concave down.

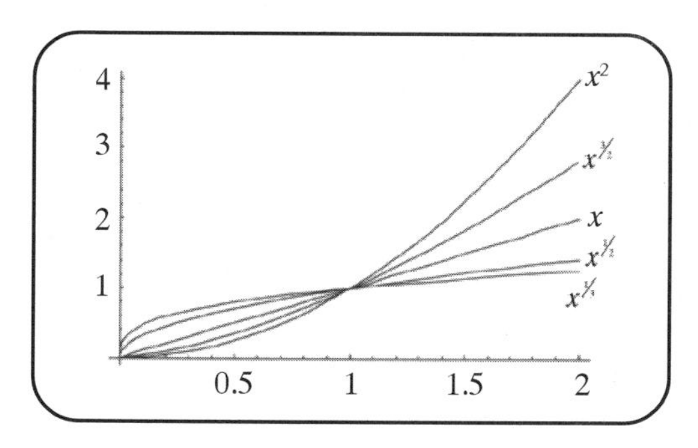

Figure 2.6.

Our discussion of power functions has been limited to power functions of the form $f(x)=x^p$. In general, a power function is a function of the form $f(x)=ax^p$, with the constant a inserted. The effect of this constant is to stretch or contract the function (or reflect the function across x-axis if $a < 0$).

When comparing power functions for large values of x, the function with the higher exponent will eventually dominate, but there is more to analyzing power functions than just looking at their asymptotic behavior.

Consider the functions $f(x)=x^3$ and $g(x)=100x^2$. Because of the difference in powers, we know that for large values of x, $f(x) > g(x)$. The question is, how large does x need to be? If $f(x) = g(x)$, then $x^3 = 100x^2$, and $x = 100$. So, for values of x greater than 100, $f(x)$ will dominate, and for values of x between 0 and 100, $g(x)$ will dominate. The effect of the constant a is that the point of intersection of power functions will change, but the constant a will not affect the long-term behavior of the power functions. When comparing the functions $h(x)=10000000x^2$ and $k(x)=0.0000000001x^5$, for large values of x the function $k(x)$ will dominate $h(x)$, even though the coefficient of $k(x)$ is very small. The coefficient will affect *how large* x has to be before $k(x)$ dominates $h(x)$. Larger powers will always eventually dominate smaller powers.

Power functions are important functions to analyze and understand. When x is large enough, *every* polynomial or rational function will behave like a power function. Polynomials and rational functions *asymptotically* approach power functions.

Lesson 2-6: Piecewise-Defined Functions

A **piecewise-defined function** is a function that is defined by different formulas on different parts of its domain. When evaluating a piecewise-defined function, you must first locate the region of the domain that contains the value of the independent variable.

One piecewise-defined function that appears frequently in mathematics is the absolute value function. The absolute value of a number can never be negative. The absolute value of a positive number is itself, and the absolute value of a negative number is the resulting positive number obtained by multiplying the negative number by −1. We can write the absolute value of a number as a function as follows:

$$|x| = \begin{cases} -x & x < 0 \\ x & x \geq 0 \end{cases}$$

Its graph is shown in Figure 2.7.

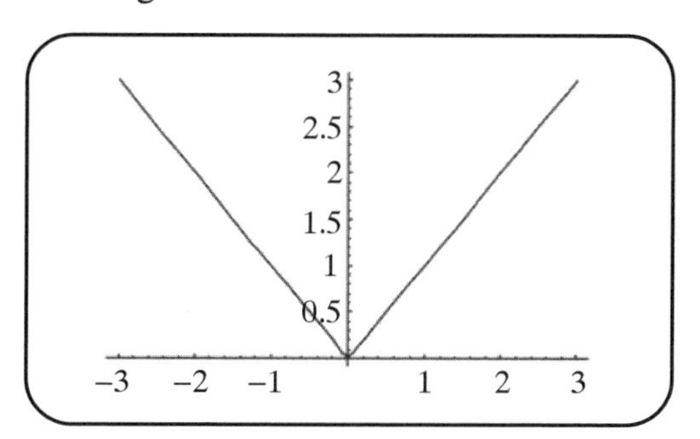

Figure 2.7.

To graph a piecewise-defined function, first break the domain into the pieces indicated by the inequalities. Graph each function in its appropriate piece, and don't extend any of the graphs beyond the boundary of the regions. Figure 2.8 is the graph of the piecewise-defined function:

$$f(x) = \begin{cases} 1 - x & x < 1 \\ x - 2 & x \geq 1 \end{cases}$$

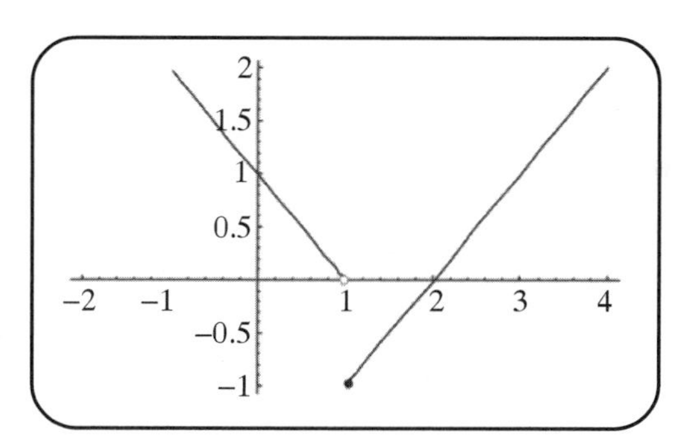

Figure 2.8.

There are some observations to make. First of all, notice that at the boundary $x = 1$, the two pieces of the function do not meet. One of the pieces ends with an open circle; the other ends with a closed circle. The side of the domain with the strict inequality has the open circle. If one of the sides is either ≥ or ≤, then its graph will end at the boundary and have a closed circle. The closed circle indicates that if a point lies on the boundary of two regions, the value of the function is determined by the rule that does *not* involve a strict inequality. If both sides involve a strict inequality, then both sides will have open circles at the boundary.

In general, the domain of a function can be broken up into many pieces, but the pieces *cannot* overlap. We cannot have two different rules apply to a particular input value. Piecewise-defined functions are often used to determine cell phone bills, utility bills, income tax, royalties on book sales, or situations in which a discount is given for a bulk purchase above a threshold amount.

Answer Key

Lesson 2-1 Review

1. $y = \frac{2}{3}x + \frac{11}{3}$

2. y-intercept: $(0,-4)$; x-intercept: $\left(\frac{4}{3},0\right)$

3. Let x represent the number of T-shirts made, and $C(x)$ represent the cost. $C(x) = 4x + 50$ and $C(500) = 2{,}050$

Lesson 2-2 Review

1. $f(x) = x^2 - 8x + 12$: concave up;
   axis of symmetry: $x = 4$;
   vertex: $(4,-4)$;
   y-intercept: $(0,12)$;
   x-intercepts: $(6,0)$ and $(2,0)$

2. $f(x) = -x^2 + 6x - 4$: concave down;
   axis of symmetry: $x = 3$;
   vertex: $(3,5)$;
   y-intercept: $(0,-4)$;
   x-intercepts: $\left(-2+\sqrt{10},0\right)$ and $\left(-2-\sqrt{10},0\right)$

3. a. $f(x) = x^2 + 2x$: axis of symmetry is $x = -\frac{2}{2(1)} = -1$.

   The parabola is concave up, so the function will be decreasing on $(-\infty,-1)$ and increasing on $(-1,\infty)$.

b. $g(x) = -3x^2 - 2x + 1$: axis of symmetry is $x = -\frac{-2}{2(-3)} = -\frac{1}{3}$.

The parabola is concave down, so the function will be increasing on $\left(-\infty, -\frac{1}{3}\right)$ and decreasing on $\left(-\frac{1}{3}, \infty\right)$.

4. This is equivalent to solving the inequality $x^2 - 2x - 15 < 0$. Factor the quadratic equation: $x^2 - 2x - 15 = (x-5)(x+3)$. Create a sign chart for the function, as shown in Figure 2.9. The inequality is satisfied in the interval $(-3, 5)$.

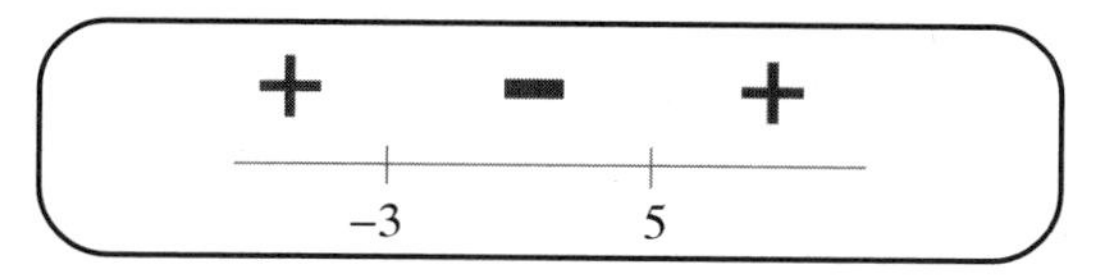

Figure 2.9.

Alternatively, the graphs of the functions $f(x) = x^2$ and $g(x) = 2x + 15$ are shown in Figure 2.10, and we can see that $f(x) < g(x)$ on the interval $(-3, 5)$.

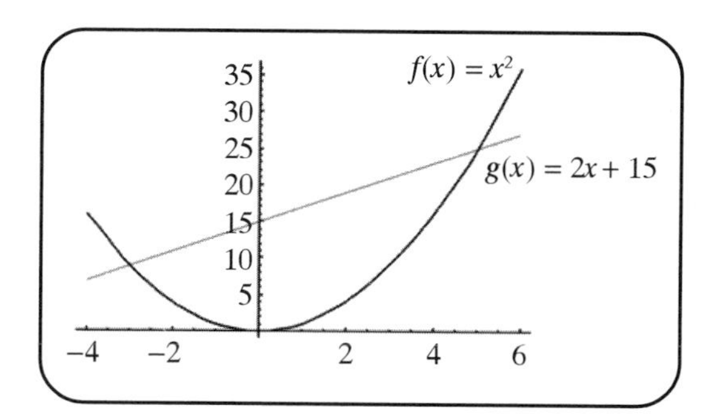

Figure 2.10.

Lesson 2-3 Review

1. a. $f(x) = 3x^4 - 2x + 1$: neither even nor odd

   b. $f(x) = 3x^{10} - 2x^6 + 5$: even

   c. $f(x) = 3x^9 - 2x^5 + 1$: neither even nor odd

   d. $f(x) = 3x^9 - 2x^5 + x$: odd

Lesson 2-4 Review

1. $G(x) = \frac{x^2+x-6}{2x^2-x-3} = \frac{(x+3)(x-2)}{(2x-3)(x+1)}$:

   y-intercept: $(0, 2)$;
   x-intercepts: $(-3, 0)$ and $(2, 0)$;

   horizontal asymptote: $y = \frac{1}{2}$;

   vertical asymptotes: $x = -1$ and $x = \frac{3}{2}$

2. $H(x) = \frac{x^2-1}{x^3-5x^2+6x} = \frac{(x-1)(x+1)}{x(x-3)(x-2)}$:

   y-intercept: none;
   x-intercepts: $(1, 0)$ and $(-1, 0)$;
   horizontal asymptote: $y = 0$;
   vertical asymptotes: $x = 0, x = 3$, and $x = 2$

1 2 3 4 5 6 7 8 9 10 11 12 13 14 15

Exponential and Logarithmic Functions

In the last chapter we learned about power functions. Power functions involve an independent variable raised to a power. Power functions, such as the function $f(x) = x^2$, are exponential expressions in which the base is the independent variable and the exponent is a constant. When we change things around so that the base is a constant and the independent variable is in the exponent, as with the function $g(x) = 2^x$, we create an exponential function. In this chapter, we will examine and transform exponential functions. We will also examine logarithmic functions, which are the inverses of exponential functions.

Lesson 3-1: Exponential Functions

An **exponential function** is a function of the form $f(x) = b^x$, where the base, b, is a positive constant. Some examples of exponential functions are $f(x) = 2^x$ and $g(x) = 10^x$. Because $a^0 = 1$ if $a \neq 0$, every exponential function passes through the point (0, 1).

If the base of an exponential function is greater than 1, the function will increase exponentially and exhibit **exponential growth.** If the base of an exponential function is between 0 and 1, we say that the function decreases exponentially and exhibits **exponential decay.**

The value of the base will affect the growth rate of the exponential function. For exponential growth, a larger base will result in a larger growth rate. The closer the base of an exponential function is to 1, the less dramatic the exponential growth or decay of the function.

An exponential function can be transformed by translation, reflection, expansion, and contraction. The graph of the function $b^x + c$ is the graph of b^x shifted up/down c units. The graph of the function b^{x+c} is the graph of b^x shifted to the left/right c units. The graph of the function $-b^x$ is the graph of b^x reflected across the x-axis, and the graph of b^{-x} is the graph of b^x reflected about the y-axis.

Reflecting an exponential function across the y-axis turns an exponential growth function into an exponential decay function, or vice versa. Shifting an exponential function horizontally is equivalent to stretching or contracting the function horizontally.

Figure 3.1 shows an exponential growth function. Every exponential function has the same shape as this function. The graph of an exponential decay function will be the reflection of an exponential growth function across the y-axis. Translating, reflecting, stretching, and contracting an exponential function will only change the scale or position of the function. These transformations will not change the shape of the function. Just as we saw with parabolas, there is not much variety when it comes to the shape of an exponential function.

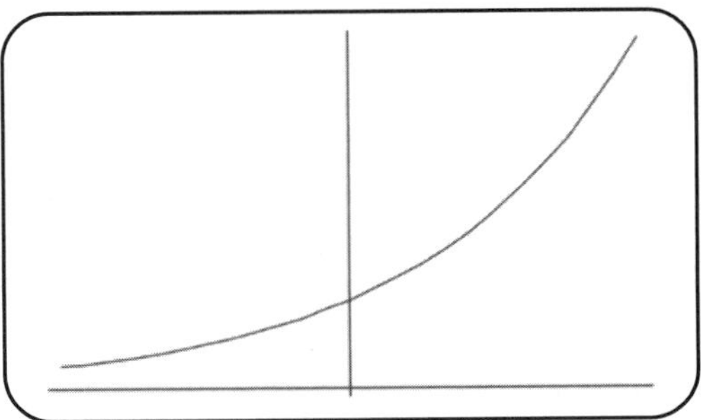

Figure 3.1.

Exponential growth functions have a one-sided horizontal asymptote. As $x \to -\infty$, an exponential growth function gets close to 0, so the horizontal asymptote of an exponential growth function is $y = 0$. Exponential decay functions also have a one-sided horizontal asymptote $y = 0$. In general, functions of the form $f(x) = a \cdot b^x$, whether they represent exponential growth or decay, only have one-sided horizontal asymptotes.

Exponential growth is thought of as being extremely rapid, and over time, or for large values of x, exponential growth will be more significant than the growth of any power function. Because polynomials and rational functions behave as power functions for large values of x, exponential growth will eventually overtake *any* polynomial, rational function, or power function. This generalization only holds for long periods of time; there are situations when polynomial growth is faster than exponential growth.

Consider the functions $f(x) = 2^x$ and $g(x) = x^2$. The graphs of these two functions are shown in Figure 3.2. Notice that between $x = 0$ and $x = 2$, the exponential function is higher than the power function. Between $x = 2$ and $x = 4$, the power function overtakes the exponential function, but for $x > 4$ the exponential function overtakes the power function permanently. *Every* exponential growth function will eventually dominate *every* power function.

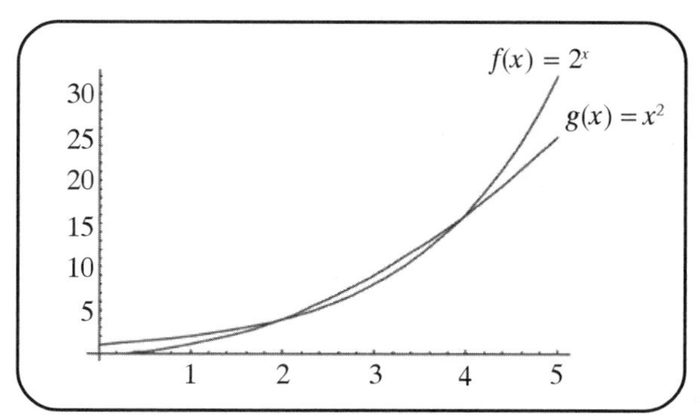

Figure 3.2.

We can also compare exponential decay with power decay. A **power decay** function is a power function that has a negative exponent. Consider the functions $f(x) = \left(\frac{1}{2}\right)^x$ and $g(x) = x^{-2}$. Both functions have the horizontal asymptote $y = 0$. As x gets large, both functions get small. Because $\left(\frac{1}{2}\right)^x = \frac{1}{2^x}$ and $x^{-2} = \frac{1}{x^2}$, the fact that 2^x will eventually dominate x^2 means that $\left(\frac{1}{2}\right)^x$ will eventually be less than x^{-2}. In fact, *every* exponential decay function will eventually approach the horizontal asymptote $y = 0$ faster than *every* power decay function.

If the function $f(x) = b^x$ is shifted vertically, then the horizontal asymptote will shift by the same amount. The graph of the function $f(x) = 2^x - 1$ is the graph of 2^x shifted down 1 unit, and the horizontal asymptote of $f(x) = 2^x - 1$ is $y = -1$. Similarly, the function $g(x) = 3^{-x} + 5$ has the horizontal asymptote $y = 5$. If an exponential function is shifted horizontally, then the horizontal asymptote will not change. A horizontal line is not affected by a horizontal translation, but it is affected by a vertical translation.

A common application of exponential models occurs when calculating compound interest. If an amount of money P, called the principal, is invested at an annual interest rate r (as a decimal) and interest is compounded n times per year, then the formula to calculate the amount of money after t years is:

$$A = P\left(1+\tfrac{r}{n}\right)^{n\cdot t}$$

where A is the amount of money after t years.

The interest earned on an investment will increase as the number of times that interest is compounded increases. It should come as no surprise that financial institutions will compound frequently on loans and credit card balances (when *you* are the one paying the interest), and compound as infrequently as possible on savings and checking accounts (when *they* are the ones paying the interest)! If we take things to an extreme and imagine that interest is being compounded continuously, the formula used to calculate A will change. In this situation, we are looking at the asymptotic behavior of A as n, the number of times we compound interest, approaches infinity.

In the equation $A = P\left(1+\tfrac{r}{n}\right)^{n\cdot t}$, if $m = \tfrac{n}{r}$, then $\tfrac{r}{n} = \tfrac{1}{m}$ and $n \cdot t = m \cdot r \cdot t$. We can substitute these expressions into our formula and simplify:

$$A = P\left(1+\tfrac{r}{n}\right)^{n\cdot t}$$

$$A = P\left(1+\tfrac{1}{m}\right)^{m\cdot r\cdot t}$$

$$A = P\left[\left(1+\tfrac{1}{m}\right)^{m}\right]^{r\cdot t}$$

Now, as the number of times that the interest is compounded heads towards infinity, so does m. And as $m \to \infty$, the expression $\left(1+\tfrac{1}{m}\right)^{m}$ does something interesting. In the expression $\left(1+\tfrac{1}{m}\right)^{m}$, both the base and the exponent are changing. The exponent m will try to make the expression exhibit exponential growth (because the base is greater than 1). But as m increases, the base gets closer to 1. And the closer to 1 the base becomes, the slower the growth of the function. These two actions are contradictory and, instead of the function growing exponentially, it will actually taper off and asymptotically approach a fixed constant. This fixed constant is an irrational number that appears in some surprising places. It is denoted by the letter e, and its numeric value is 2.71828....

This isn't the first time that mathematicians have referred to special numbers by a letter: Pi (π) is a letter of the Greek alphabet used to represent the ratio of the circumference to the diameter of any circle, and its numeric value is 3.141592653589.... Pi has its origins in geometry, and e has its origins in calculus. Analyzing the expression $\left(1+\frac{1}{m}\right)^m$ as $m \to \infty$ is our first glimpse into the world of calculus. This analysis is referred to as evaluating a limit, and we will see this expression again in Chapter 12.

We can interpret e using the ideas of compound interest. If you deposit \$1 in an account that earns interest at an annual rate of 100% compounded continuously, the amount of money in the account after one year will be \$$e$. The function $f(x) = e^x$ appears throughout science, engineering, and finance. It is an exponential function with a base larger than 1, so the function will exhibit exponential growth.

To understand the connection between $f(x) = e^x$ and finance, we will continue with our original investigation. We were analyzing the function:

$$A = P\left[\left(1+\tfrac{1}{m}\right)^m\right]^{r\cdot t} \quad \text{as } m \to \infty$$

We can use the fact that $\left(1+\frac{1}{m}\right)^m \to e$ as $m \to \infty$ to obtain the formula:

$$A = Pe^n$$

This formula can be used to calculate the amount of money, A, after t years if P dollars is invested at an annual interest rate r (written as a decimal) compounded continuously. Keep in mind that compounding continuously is as good as it gets.

Lesson 3-2: Inverse Functions

A function is used to represent the relationship, or dependence, between one quantity and another. The rate at which a cricket chirps is related to, or is a function of, the temperature of its environment. Under certain circumstances, we can use a cricket as a thermometer. In this situation, we are looking at the environmental temperature as a function of the chirp rate. Looking at things from the reverse perspective can be thought of mathematically as inverting a function.

There are rules for how to invert a function, and there are conditions under which a function *cannot* be inverted. We denote the inverse of the function f by f^{-1}. When dealing with rational and exponential expressions, we have interpreted the superscript –1 to denote the reciprocal of a number. Now we are using it to denote the inverse of a function. There is a reason that we are reusing this notation. The *reciprocal* of a number a is the number to *multiply* a by to get 1 (the *multiplicative identity*). The **inverse** of a function f is a function that you *compose* with f to get the *identity function*.

An identity is a process that does not change the input. For example, the *additive* identity is the number that does not change any number under addition: The additive identity is 0. Similarly, the *multiplicative* identity is the number that does not change any number under multiplication: The multiplicative identity is 1. The **identity function** is the function that does not change any input value of the function: The identity function is $f(x) = x$. Notice that the input of the function is the same as the output of the function. The inverse of a function f is the unique function, denoted f^{-1}, that has the following relationship to f:

$$\left(f^{-1} \circ f\right)(x) = x \text{ and } \left(f \circ f^{-1}\right)(x) = x$$

If a function has an inverse, the function is **invertible**, and the inverse of an invertible function is unique.

As I mentioned earlier, the inverse of a function involves a change in perspective: The output becomes the input, and the input becomes the output. If f is an invertible function with domain X and range Y, then f^{-1} will be a function with domain Y and range X. When inverting a function, the role of the independent variable, x, and the role of the dependent variable, y or f, switch. This concept will be useful in actually finding the inverse of invertible functions. If an invertible function f passes through the point (a, b), then its inverse, f^{-1}, will pass through the point (b, a): The input becomes the output, and the output becomes the input. If f represents the chirp rate of crickets as a function of temperature, where temperature is measured in degrees Fahrenheit, and $f(60) = 100$, then this means that at a temperature of 60° F, the chirp rate is 100 chirps per minute. From this, we know that $f^{-1}(100) = 60$, meaning that a chirp rate of 100 chirps per minute corresponds to an environmental temperature of 60° F. If a function f is invertible and $f(a) = b$, then $f^{-1}(b) = a$. In other words, if f is invertible, then $f^{-1}(r) = s$ means the same thing as $f(s) = r$.

Not all functions have inverses. Suppose that $f(2) = 4$ and $f(-2) = 4$. In this situation there is no unique value for $f^{-1}(4)$. Some people may decide that $f^{-1}(4) = 2$ and other people may prefer that $f^{-1}(4) = -2$. Because $f^{-1}(4)$ is not unique, we avoid problems by declaring that f is not invertible.

Our understanding of symmetry allows us to conclude that an even function is not invertible on its entire domain. Remember that an even function satisfies the relationship $f(-x) = f(x)$, so if $f(a) = b$, then $f(-a) = b$ and there is no unique choice for $f^{-1}(b)$. On the other hand, every linear function is invertible.

A function will be invertible if it is one-to-one. A function is one-to-one if no two elements in the domain have the same image. In other words, if a and b are in the domain of f, and $a \neq b$, then $f(a) \neq f(b)$. Another way to look at this situation is that if $f(a)$ and $f(b)$ are two points in the range with $f(a) = f(b)$, then $a = b$. Determining whether a function has an inverse is equivalent to determining whether the function is one-to-one.

If a function is not invertible, we may be able to restrict the domain so that it *will* be invertible. For example, the function $f(x) = x^2$ is not one-to-one (because it is an even function), so it is not invertible. If we restrict the domain of f to be the non-negative real numbers, then it will be invertible. With this restriction, power functions will have inverses: The inverse of $f(x) = x^p$ will be $f^{-1}(x) = x^{1/p}$. The inverse of a power function is a root. We use this relationship without much thought in algebra. In order to solve the equation $x^2 = 4$, most people simply "take the square root of both sides and throw in a $\pm$." This process actually involves restricting the domain of the function $f(x) = x^2$ to the non-negative real numbers, and using its inverse function $f^{-1}(x) = \sqrt{x}$ to find one value of x that satisfies the equation. Then the other solution to the equation is found from symmetry; $f(x) = x^2$ is an even function so the other solution will be on the other side of the y-axis, hence the $\pm$. There is no need to change *how* you solve a quadratic equation, but it's always nice to be aware of the reason *why* you get the correct answer when you take short cuts.

If a function is both increasing and decreasing on an interval I, then it cannot be one-to-one on that interval. If a function is not one-to-one on an interval, then it cannot have an inverse on that interval. The domain of the function would have to be restricted so that only one type of behavior, either increasing or decreasing, is exhibited.

If an invertible function is defined by a formula, it is sometimes possible to find a formula for the inverse function. The key to finding a formula for the inverse of a function is to switch the role of the independent variable and the role of the dependent variable. In other words, the role of x and the role of y switch when you invert a function. To find the inverse of a function $y = f(x)$, first switch x and y, and then solve for y. To switch x and y, everywhere you see an x, write a y, and everywhere you see a y, replace it with an x.

Example 1

Find the inverse of the function $f(x) = 2x + 1$.

Solution: The function $f(x) = 2x + 1$ is a linear function, so it will have an inverse. Rewrite the function as $y = 2x + 1$. Switch x and y and then solve for y: $y = 2x + 1$ becomes $x = 2y + 1$. Now we can solve for y:

$$x = 2y + 1$$
$$2y = x - 1$$
$$y = \frac{x-1}{2}$$

Thus $f^{-1}(x) = \frac{1}{2}(x-1)$.

Notice that the function $f(x) = 2x + 1$ takes the input, doubles it, and then adds 1. The inverse function, f^{-1}, will just undo what the function f does, but in reverse order. First, f^{-1} undoes the addition of 1 by subtracting 1. Then f^{-1} undoes the multiplication by 2, by dividing by 2. You can see this in the formula for f^{-1}: $f^{-1}(x) = \frac{1}{2}(x-1)$.

Being able to actually solve for the inverse of a function can be difficult, and it may require the definition of a new function (as in the case with exponential functions), but if you have the graph of a function, there is a way to graph its inverse.

Consider the function $f(x) = \frac{1}{x+2}$. This function has the horizontal asymptote $y = 0$ and the vertical asymptote $x = -2$. The inverse of this function is $f^{-1}(x) = \frac{1}{x} - 2$. Notice that $f^{-1}(x) = \frac{1}{x} - 2$ has the vertical asymptote $x = 0$ and the horizontal asymptote $y = -2$. In general, if an invertible function has a *horizontal* asymptote of $y = a$, then its inverse will have a *vertical* asymptote of $x = a$. Similarly, if a function has a *vertical* asymptote of $x = a$, then its inverse will have a *horizontal* asymptote of $y = a$.

In general, if a function passes through the point (a, b), then its inverse will pass through the point (b, a). The only points that a function has in common with its inverse will be the points that do not change when x and y are switched. These are the points where $y = x$, or $f(x) = x$. Consider the function $f(x) = x^3$ and its inverse, $f^{-1}(x) = \sqrt[3]{x}$. The graphs of these two functions are shown in Figure 3.3. These graphs are the reflections of each other about the line $y = x$. In general, to graph of the inverse of a function f, start with the graph of f and reflect the graph about the line $y = x$.

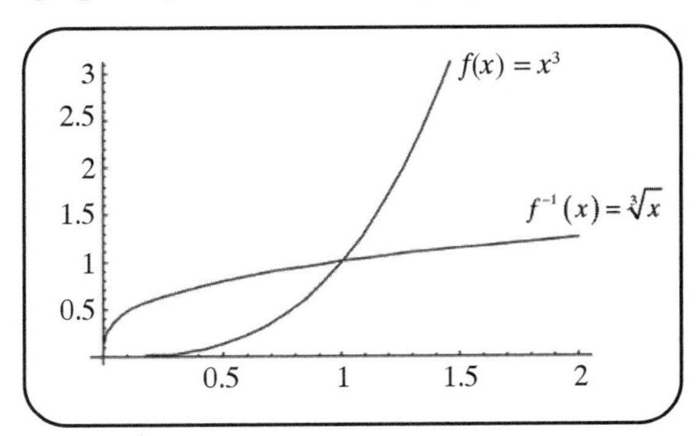

Figure 3.3.

Lesson 3-2 Review

1. Are the following functions invertible?
   a. $g(x) = x^6 + 2x^2 + 1$
   b. $h(x) = 3x + 1$

2. Find the inverse of the following functions:
   a. $f(x) = 3x - 4$
   b. $g(x) = \frac{2}{x-3}$

Lesson 3-3: Logarithmic Functions

The exponential function $f(x) = b^x$ is a one-to-one function, so it is invertible. To find its inverse, switch x and y, and then solve for y:

$y = b^x$

$x = b^y$

At this point, we are stuck. We don't know how to isolate the exponent from the base. We will define the **logarithmic function** to be the inverse of the exponential function, and write:

$$y = \log_b x \quad \leftrightarrow \quad x = b^y$$

In the *exponential* function $f(x) = b^x$, b is the base; in the logarithmic function $y = \log_b x$, b is the base of the *logarithm*. These functions are inverses of each other, and $y = \log_b x$ *means the same thing as* $b^y = x$. In other words, $x = b^y$ means that y is the power you raise b to in order to get x. Because $y = \log_b x$ means the same thing as $x = b^y$, the equation $y = \log_b x$ *also* means that y is the power you raise b to in order to get x.

For example, to evaluate $x = \log_2 8$, x is the power you raise 2 to in order to get 8: $2^x = 8$. We know that $2^3 = 8$, so $x = 3$; 3 is the power you raise 2 to in order to get 8: $\log_2 8 = 3$. To evaluate $x = \log_2 2$, think of the power that 2 must be raised to in order to get 2: $2^1 = 2$, so 1 is the power that 2 must be raised to in order to get 2: $\log_2 2 = 1$. Using this reasoning, we see that $\log_b b = 1$ for any base b.

To understand the properties of the logarithmic function, it is best to start with the exponential function. The domain of the exponential function is all real numbers, and the range is the set of positive real numbers. When you invert a function, the domain becomes the range and the range becomes the domain. Because the logarithmic function is the inverse of the exponential function, the domain of the logarithmic function is the set of positive real numbers, and the range is the set of all real numbers.

Every exponential function $f(x) = a^x$ has a y-intercept of (0, 1). From this, we know that every logarithmic function $f(x) = \log_b x$ has an x-intercept of (1, 0). Some important rules for simplifying logarithms are summarized here. These rules stem from the rules for simplifying exponents. These properties are valid only when A and B are in the domain of the logarithmic function. In other words, these properties are only true if A and B are positive real numbers.

- $\log_b(AB) = \log_b(A) + \log_b(B)$
- $\log_b\left(\frac{A}{B}\right) = \log_b(A) - \log_b(B)$
- $\log_b(A^p) = p \cdot \log_b(A)$
- $\log_b(1) = 0$
- $\log_b(b^p) = p$
- $b^{\log_b A} = A$

There are two important bases that are often used in logarithms: base 10 (referred to as the **common logarithm**) and base ***e*** (referred to as the **natural logarithm**). Technically, we can work with any base, but common logarithms are used in science, and natural logarithms appear throughout science and mathematics. Numerical values for expressions that involve common and natural logarithms can be evaluated using a standard calculator. If the base of a logarithm is not 10 or ***e***, then the base of a logarithm can be changed using the change of base formula:

$$\log_a x = \frac{\log_b x}{\log_b a}$$

If you need to evaluate the expression $\log_6 3$, it is a simple matter of evaluating $\frac{\log_{10} 3}{\log_{10} 6}$. You would get the same result if you evaluated $\frac{\log_e 3}{\log_e 6}$. The natural logarithmic function is written $\ln x$ and the common logarithmic function is written $\log x$. If the base of the logarithm is other than 10 or e, then the base must be included in the logarithm expression.

The inverse of an exponential function will be a logarithmic function (and the inverse of a logarithmic function will be an exponential function). Transforming an exponential function will transform its inverse as well.

To find the inverse of the function $f(x) = 2^x - 1$, apply the same strategy we applied earlier: Switch x and y, and then solve for y:

	$y = 2^x - 1$
Switch x and y	$x = 2^y - 1$
Add 1 to both sides of the equation	$2^y = x + 1$
Write the equation in logarithmic form	$y = \log_2 (x + 1)$

The function $f(x) = 2^x - 1$ has the horizontal asymptote $y = -1$, so its inverse, $f^{-1}(x) = \log_2 (x + 1)$, will have the vertical asymptote $x = -1$. The graph of $f^{-1}(x) = \log_2 (x + 1)$ will be the reflection of the graph of $f(x) = 2^x - 1$ about the line $y = x$. The graphs of $f(x) = 2^x - 1$ and $f^{-1}(x) = \log_2 (x + 1)$ are shown in Figure 3.4 on page 58.

Comparing the graphs of $f(x) = 2^x - 1$ and $f^{-1}(x) = \log_2 (x + 1)$, we see that exponential growth is much more rapid than logarithmic growth. In fact, exponential growth will *always* eventually dominate logarithmic growth. Moreover, logarithmic growth is gradual enough that *any* power function with a positive exponent, no matter how small the exponent is, will *eventually* dominate every logarithmic function. The function $f(x) = x^{0.0000001}$ will eventually overtake $10000000 \log_2 x$. It may take a while, but it *will* happen.

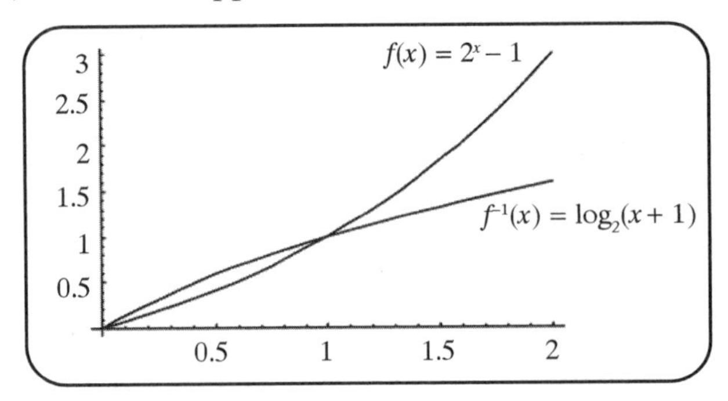

Figure 3.4.

Lesson 3-4: Exponential and Logarithmic Equations

Solving exponential and logarithmic equations involves similar steps. First, isolate the exponential or logarithmic expressions on one side of the equation. If the equation involves solving for the exponent, take the logarithm of both sides (either the common logarithm or the natural logarithm) and use the properties of logarithms to simplify the new equation. If the equation involves solving for the argument of a logarithm function, rewrite the equation in exponential form and then simplify.

We will start by solving exponential equations. An **exponential** equation is an equation in which the variable occurs in the exponent.

Example 1

Solve the equation $3^x = 8$ to six decimal places.

Solution: The exponential expression is already isolated, so we'll take the natural logarithm of both sides and then solve for x:

	$3^x = 8$
Take the natural logarithm of both sides	$\ln(3^x) = \ln 8$
Use the properties of logarithms to simplify	$x \cdot \ln 3 = \ln 8$
Divide both sides by ln 3	$x = \frac{\ln 8}{\ln 3} = \frac{2.07944}{1.09861}$ $= 1.89279$

Example 2

Solve the equation $4^{2x+1} = 9$ to six decimal places.

Solution: The exponential expression is already isolated, so we'll take the common logarithm of both sides and then solve for x:

	$4^{2x+1} = 9$
Take the common logarithm of both sides	$\log(4^{2x+1}) = \log 9$
Use the properties of logarithms to simplify	$(2x+1)\log 4 = \log 9$
Divide both sides by log 4	$(2x+1) = \frac{\log 9}{\log 4}$
Subtract 1 from both sides	$2x = \frac{\log 9}{\log 4} - 1$
Divide both sides by 2	$x = \frac{1}{2}\left(\frac{\log 9}{\log 4} - 1\right)$
Use a calculator to evaluate the expression	$x = 0.292481$

Example 3

Solve the equation $x^2(2^x) - 6x(2^x) = 0$.

Solution: Both terms in this equation have an exponential component 2^x in common, so we can factor it out: $x^2(2^x) - 6x(2^x) = (2^x)(x^2 - 6x)$. Now we can solve the equation $(2^x)(x^2 - 6x) = 0$. Because $2^x \neq 0$ for all x, the only way for this product to equal 0 is if the polynomial component is equal to 0: $(x^2 - 6x) = 0$. This equation can be solved by factoring:

$$x(x-6) = 0$$

$$x = 0 \qquad\qquad x - 6 = 0$$

$$x = 6$$

The solutions are $x = 0$ and $x = 6$.

Example 4

Solve the equation $e^{2x} - 3e^x - 4 = 0$.

Solution: This problem doesn't seem to fit into the same category as the other problems, but with a simple substitution, it will. If we let $t = e^x$, then $t^2 = (e^x)^2 = e^{2x}$, and the equation to solve becomes $t^2 - 3t - 4 = 0$. This is a quadratic equation that can be solved by factoring:

$$t^2 - 3t - 4 = 0$$

$$(t-4)(t+1) = 0$$

$$t = 4 \qquad t = -1$$

Now that we know the values for t, we can solve for x by using the substitution equation $t = e^x$ when $t = 4$:

	$t = e^x$
Substitute $t = 4$ into the equation	$e^x = 4$
Take the natural logarithm of both sides	$\ln(e^x) = \ln 4$
Use the property of logarithms	$x \cdot \ln e = \ln 4$
$\ln e = 1$	$x = \ln 4$
Now we can examine what happens if $t = -1$:	$t = e^x$
	$e^x = -1$

This equation has no solution, because $e^x > 0$ for all x.

The solution to the equation $e^{2x} - 3e^x - 4 = 0$ is $x = \ln 4$.

We will now turn our attention to solving logarithmic equations. A **logarithmic** equation is an equation in which the variable occurs in the argument of a logarithm function. To solve these equations, isolate the terms that involve logarithms on one side of the equation, and move all non-logarithm terms to the other side. If there is more than one term that involves a logarithm, use the properties of logarithms to combine them. Then write the equivalent equation in exponential form by using the relationship

$$y = \log_b x \qquad \leftrightarrow \qquad x = b^y$$

Finally, solve the resulting equation. If you use the properties of logarithms to combine two or more logarithmic expressions, you *must* check your answers in the *original* problem. When you combine two or more logarithmic expressions, it is possible to introduce extraneous solutions that must later be rejected. An **extraneous solution** is a solution to an intermediate equation that is not a solution of the original equation. Extraneous solutions can be introduced when you combine two logarithmic expressions, when you square both sides of an equation, or whenever you perform an operation that is not invertible.

Example 5

Solve the equation $\log_3 (x - 1) = 2$.

Solution: There is only one term that involves a logarithm, and it is already isolated, so we can write the equivalent equation in exponential form and solve it:

$$\log_3 (x - 1) = 2 \quad \leftrightarrow \quad 3^2 = (x - 1)$$

$$9 = (x - 1)$$

$$x = 10$$

Example 6

Solve the equation $\ln (x + 1) = 1$.

Solution: There is only one term that involves a logarithm, and it is already isolated, so we can write the equivalent equation in exponential form and solve it:

$$\ln (x + 1) = 1 \quad \leftrightarrow \quad e^1 = (x + 1)$$

$$e = (x + 1)$$

$$x = e - 1$$

Example 7

Solve the equation $\log_2 x + \log_2 (x + 2) = 3$.

Solution: Use the properties of logarithms to combine the two logarithmic expressions:

$$\log_2 x + \log_2 (x+2) = \log_2 \left[x(x+2)\right]$$

Now, solve the equation $\log_2 \left[x(x+2)\right] = 3$ by writing the equivalent equation in exponential form:

$$\log_2 \left[x(x+2)\right] = 3 \quad \leftrightarrow \quad 2^3 = \left[x(x+2)\right]$$

Now, solve the equation $2^3 = \left[x(x+2)\right]$:

	$2^3 = \left[x(x+2)\right]$
Expand both expressions	$8 = x^2 + 2x$
Subtract 8 from both sides	$x^2 + 2x - 8 = 0$
Factor	$(x+4)(x-2) = 0$
Set each factor equal to 0 and solve	$x = -4$ or $x = 2$

Because we used the properties of logarithms to combine two logarithmic expressions, we must make sure that the values of x that satisfy the quadratic equation $2^3 = [x(x+2)]$ make sense in the *original* equation. If one or both of the values of x do not make sense in the original equation, then we must discard those values of x, as they will be extraneous solutions. Substituting $x = -4$ into the original equation yields $\log_2(-4) + \log_2(-4+2)$, and these expressions do not make sense because the domain of the logarithm function is the set of positive real numbers. So $x = -4$ is an extraneous solution. Substituting $x = 2$ into the original equation yields $\log_2(2) + \log_2(2+2)$. Both of these expressions make sense, meaning that $x = 2$ is a valid solution to the original equation. Therefore, the only solution to the equation $\log_2 x + \log_2(x+2) = 3$ is $x = 2$.

Lesson 3-4 Review

Solve the following exponential and logarithmic equations.

1. $3^x = 4^{2x-1}$
2. $x^2(3^x) - 8x(3^x) + 12(3^x) = 0$
3. $e^{4x} - 5e^{2x} + 4 = 0$
4. $\log_4(1-x) = 1$
5. $\ln(3x+2) = 2$

6. $\log_{20}(x+3)+\log_{20}(x+2)=1$

7. $\log_3(x+1)-\log_3(x-1)=1$

Answer Key

Lesson 3-2 Review

1. a. $g(x)=x^6+2x^2+1$: not invertible, because it is an even function

   b. $h(x)=3x+1$: invertible, because it is a linear function

2. Find the inverse of the following functions:

   a. $f(x)=3x-4$: $f^{-1}(x)=\frac{1}{3}(x+4)$

   b. $g(x)=\frac{2}{x-3}$: $g^{-1}(x)=\frac{2}{x}+3$

Lesson 3-4 Review

1. $\ln(3^x)=\ln(4^{2x-1})$

   $x\ln 3=(2x-1)\ln 4$

   $x(1.10)=(2x-1)(1.39)$

   $x\approx 0.83$

2. $(3^x)(x^2-8x+12)=0$

   $(3^x)(x-6)(x-2)=0$

   $x=6$ or $x=2$

3. Let $t=e^{2x}$:

   $t^2-5t+4=0$

   $(t-4)(t-1)=0$

   $t=4$ or $t=1$

   $e^{2x}=4$ or $e^{2x}=1$

   $x=\ln 2\approx 0.693$ or $x=0$

4. Write the equation in exponential form and solve:

$$(1-x)=4^1$$

$$x=-3$$

5. Write the equation in exponential form and solve:

$$(3x+2)=e^2$$

$$x=\frac{1}{3}(e^2-2)$$

6. Combine the two terms that involve logarithms, write the resulting equation in exponential form, and solve:

$$\log_{20}(x+3)(x+2)=1$$

$$(x+3)(x+2)=20^1$$

$$x^2+5x+6=20$$

$$x^2+5x-14=0$$

$$(x+7)(x-2)=0$$

$x=-7$ or $x=2$

Check each answer in the original equation:

$\log_{20}(-7+3)+\log_{20}(-7+2)=1$: nonsense

$\log_{20}(2+3)+\log_{20}(2+2)=1$: no problem

There is only one solution: $x=2$; $x=-7$ is an extraneous solution.

7. Combine the two logarithm terms, write the equation in exponential form, and solve:

$$\log_3\frac{(x+1)}{(x-1)}=1$$

$$\frac{(x+1)}{(x-1)}=3^1$$

$$(x+1)=3(x-1)$$

$$x=2$$

Check the solution to make sure it is not an extraneous solution:

$\log_3(2+1)-\log_3(2-1)=1$: no problem

The solution is $x=2$.

Trigonometric Functions

The development of trigonometry stems from the study of triangles. There are two traditional approaches to defining trigonometric functions. They can either be defined in terms of the relationships between the measures of the angles and the lengths of the sides of a right triangle, or they can be defined using the unit circle. Both approaches lead to the same functions, but each approach has its advantages when solving certain types of problems. Trigonometric functions are particularly useful in describing behavior that repeats in cycles.

Lesson 4-1: Sine and Cosine

In the triangle approach to trigonometry, we start with a right triangle, as shown in Figure 4.1. We will define the sine and cosine function for acute angles. Suppose that the lengths of the two legs of the right triangle are a and b respectively, and that the length of the hypotenuse is c. By the Pythagorean theorem, we have the relationship $a^2 + b^2 = c^2$. If the side with length b is *opposite* the angle θ in the triangle, and the side with length a is *adjacent* to the angle θ, then we define $\sin\theta = \frac{b}{c}$ and $\cos\theta = \frac{a}{c}$. Right triangles are not always oriented in this way. It is beneficial to understand how to calculate the sine and cosine of an angle for any triangle orientation. Conceptually, $\sin\theta = \frac{\text{opposite}}{\text{hypotenuse}}$, and $\cos\theta = \frac{\text{adjacent}}{\text{hypotenuse}}$. From the Pythagorean theorem, we see that $a \leq c$ and $b \leq c$, which means that $\sin\theta \leq 1$ and $\cos\theta \leq 1$.

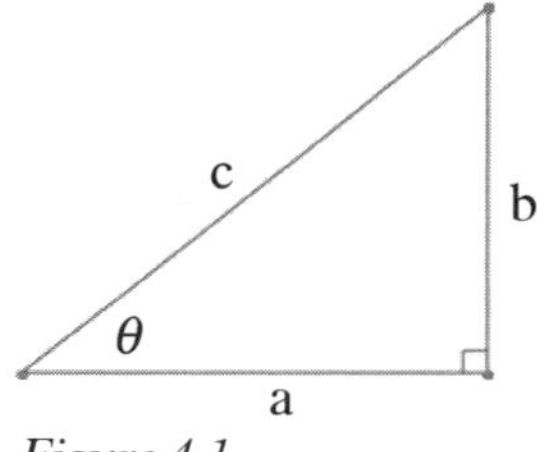

Figure 4.1.

The two most common units for the measurement of an angle are degrees and radians. Both of these units are defined in terms of the circumference of a circle, and are related by the proportion $\frac{\theta_{rad}}{\theta_{\deg}} = \frac{\pi}{180}$. Radians are more useful in calculus because the formulas are cleaner. Because of the relationship between radians and arc length, radians are particularly useful in calculating arc length. If a circle has a radius r and the arc cuts off an angle θ, as shown in Figure 4.2, then the arc length, s, can be found by the formula $s = r\theta$.

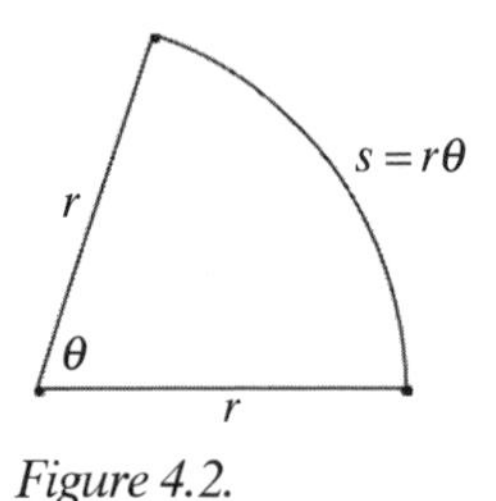

Figure 4.2.

There are two special triangles that you should know well: an isosceles right triangle and a 30° – 60° – 90° triangle. These triangles enable us to evaluate the sine and cosine of some special angles. The angles and the value of the sine and cosine of these angles are summarized in the table on this page. We will add to this table as we learn about the other trigonometric functions.

θ	$\sin\theta$	$\cos\theta$
0	0	1
$\frac{\pi}{6}$	$\frac{1}{2}$	$\frac{\sqrt{3}}{2}$
$\frac{\pi}{4}$	$\frac{\sqrt{2}}{2}$	$\frac{\sqrt{2}}{2}$
$\frac{\pi}{3}$	$\frac{\sqrt{3}}{2}$	$\frac{1}{2}$
$\frac{\pi}{2}$	1	0

Finding the sine and cosine of an angle using the properties of right triangles has one drawback: We can only calculate the sine and cosine of acute angles. If we want to expand our notion of the sine and cosine functions beyond right triangles, we will need to look at these functions through a unit circle approach.

We can define the sine and cosine of an angle of *any* measure using a unit circle. We will assume that angle measurements are always in radians, unless otherwise stated. By convention, an angle of θ radians is measured *counterclockwise* around the circle starting at the point (1, 0), as shown in Figure 4.3 on page 67. In other words, one of the rays of the angle θ is *always* the x-axis. Suppose that the intersection of the unit circle with the other ray that forms the angle θ is the point P, whose coordinates are (x, y). We *define* $\sin\theta = y$ and $\cos\theta = x$. As P moves around the unit circle, the values of $\sin\theta$ and $\cos\theta$ will oscillate between 1 and –1. Negative angles are interpreted as the angle θ moving in the *clockwise* direction.

The domain of the sine and cosine functions, as defined using the unit circle, is the set of all real numbers. From the unit circle definition of the sine and cosine functions:

$\sin\theta = y$

$\cos\theta = x$,

we can see that $\sin \theta > 0$ if the y-coordinate of the point P is positive. In other words, if the point P lies in quadrants I or II, the sine of the angle formed will be positive. If the point P lies in quadrants III or IV, then the sine of the angle formed will be negative. Similarly, $\cos \theta > 0$ if the x-coordinate of the point P is positive. In other words, if the point P lies in quadrants I or IV, the cosine of the angle formed will be positive, and if the point P lies in quadrants II or III, the cosine of the angle formed will be negative. The signs of the sine and the cosine functions for the four quadrants are listed in the following table.

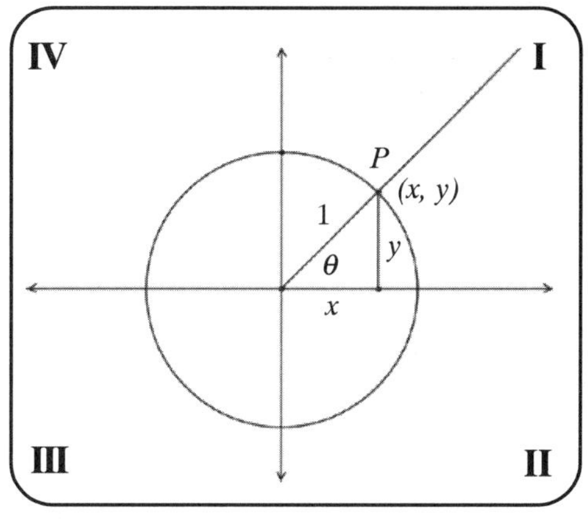

Figure 4.3.

Quadrant	Sign of $\sin \theta$	Sign of $\cos \theta$
I	+	+
II	+	−
III	−	−
IV	−	+

For acute angles, the definitions of the sine and cosine functions using a right triangle are the same as the definitions based on the unit circle approach.

Defining the sine and cosine of an angle using the unit circle enables us to create and analyze the sine and cosine *functions*. The graphs of these functions will shed some light on their nature. The sine and cosine functions are **cyclic**, or periodic functions. A **periodic function** is a function that repeats itself after a certain, fixed amount has been added to the independent variable. This property can be represented algebraically as: $f(x+p) = f(x)$. The fixed constant p is called the period of the function. The **period** of a periodic function is the time needed for the function to complete one cycle. Using the unit circle as the basis for the definition of the sine and cosine functions, we can visualize the periodic nature of these functions. After going around the circle once, the values of the sine and cosine functions start to repeat. Each time the circle is traced out, 2π radians are added to the angle, so the period of the sine and cosine functions is 2π.

The **amplitude** of a periodic function is one-half the distance between the maximum and minimum values of the function. Because the sine and cosine functions oscillate between 1 and –1, the amplitude of the sine and cosine functions is 1: $\frac{1-(-1)}{2}$.

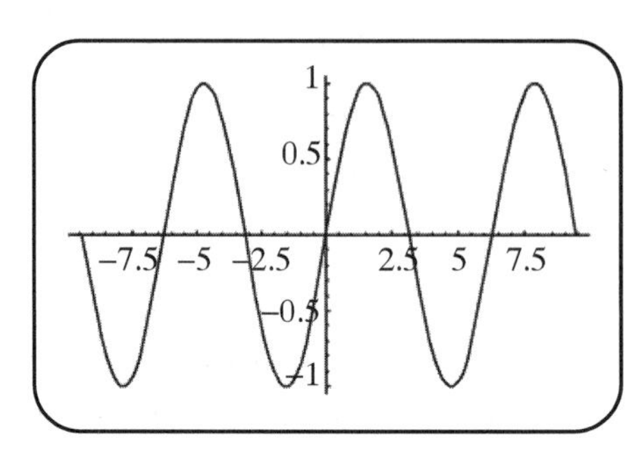

Figure 4.4.

We can use the values of some of the special angles to get a rough sketch of the graph of the sine function. If we had the patience to plot enough points, the result would be the graph shown in Figure 4.4. The sine wave, as it is sometimes called, continues indefinitely to the left and the right, and it repeats the cycle every 2π radians. The graph of the sine function will never be higher than 1 and never lower than –1. The sine function satisfies the inequality $-1 \leq \sin x \leq 1$. The domain of the sine function is the set of all real numbers. The sine function is symmetric about the origin; $f(x) = \sin x$ is an odd function, and $\sin(-x) = -\sin x$.

The graph of the cosine function is the graph of the sine function shifted to the left by $\frac{\pi}{2}$ units. The graph of the cosine function is shown in Figure 4.5. The cosine function also satisfies the inequality $-1 \leq \cos x \leq 1$. The cosine function is symmetric about the y-axis; $f(x) = \cos x$ is an even function, and $\cos(-x) = \cos x$.

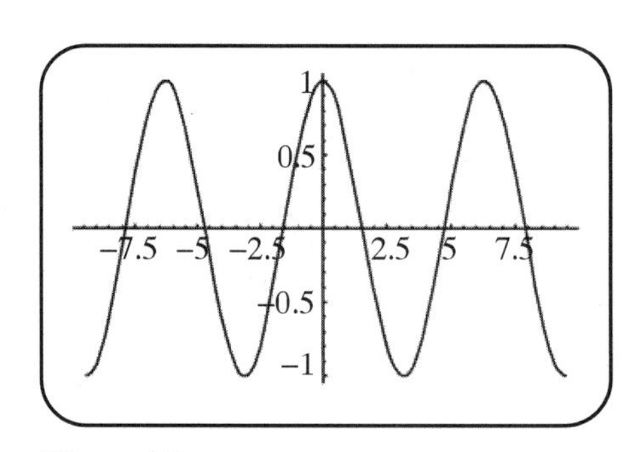

Figure 4.5.

From the graphs of the sine and cosine functions, or from the periodic nature of these functions, we can conclude that these functions have infinitely many roots, or zeros. It is important to be able to describe the location of all of the zeros of the sine and cosine functions. If n is any integer, $\sin(\pi n) = 0$ and $\cos(\pi n) = (-1)^n$. The cosine function enables us to represent an alternating sign (positive to negative) in a compact way.

We can also find the roots of the cosine function. The cosine of an angle that is an odd half-integer multiple of π is 0. If n is any positive integer, then $2n + 1$ will be an odd number, and $\cos\left(\frac{(2n+1)\pi}{2}\right)=0$. The sine of an angle that is a half-integer multiple of π is either 1 or −1: $\sin\left(\frac{(2n+1)\pi}{2}\right)=(-1)^n$. The pattern for finding the roots of the sine and cosine function is very important in analyzing the other trigonometric functions.

Lesson 4-2: The Tangent Function

The sine and cosine functions are the building blocks of the other four trigonometric functions. The tangent of an angle is the ratio of the sine to the cosine of the angle:

$$\tan\theta = \frac{\sin\theta}{\cos\theta}$$

θ	$\sin\theta$	$\cos\theta$	$\tan\theta$
0	0	1	0
$\frac{\pi}{6}$	$\frac{1}{2}$	$\frac{\sqrt{3}}{2}$	$\frac{\sqrt{3}}{3}$
$\frac{\pi}{4}$	$\frac{\sqrt{2}}{2}$	$\frac{\sqrt{2}}{2}$	1
$\frac{\pi}{3}$	$\frac{\sqrt{3}}{2}$	$\frac{1}{2}$	$\sqrt{3}$
$\frac{\pi}{2}$	1	0	undefined

The special angles that we studied in the last lesson continue to be important. I will include the values of the tangent function for these angles in the table we started in the last lesson.

Before we can graph the tangent function, we must determine its domain. The tangent function is defined as $\tan x = \frac{\sin x}{\cos x}$, and its domain will be all values of x for which $\cos x \neq 0$. The tangent function will have vertical asymptotes at all of the odd half-integer multiples of π. The tangent function has infinitely many roots, and they are located at the integer multiples of π.

The tangent function will be positive where the sign of the sine and cosine functions are the same. Recall the table of the signs of the sine and cosine functions discussed earlier. The sine and cosine functions are both positive in quadrant I and both negative in quadrant III. The sine and cosine functions have opposite signs in quadrants II and IV. The graph of the tangent function is shown in Figure 4.6 on page 70.

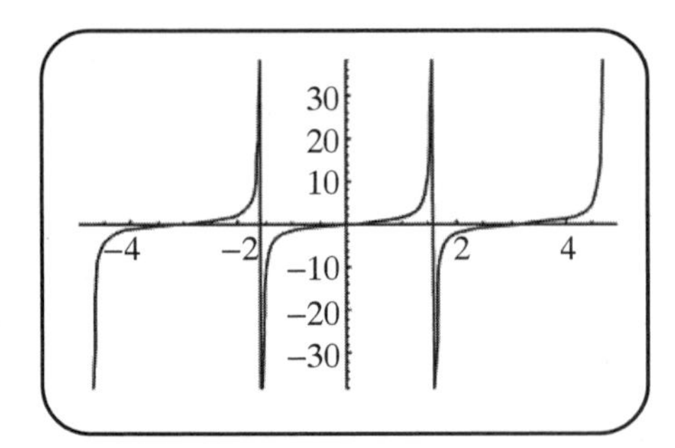

Figure 4.6.

The graph of the tangent function is symmetric about the origin; the tangent function is an odd function:

$$\tan(-x) = \frac{\sin(-x)}{\cos(-x)} = \frac{-\sin x}{\cos x} = -\tan x$$

The period of the tangent function is π rather than 2π. The tangent function is not bounded like the sine and cosine functions; the range of the tangent function is the set of all real numbers. The tangent function is periodic, but it does not oscillate like the sine and cosine functions. The other trigonometric functions will behave similarly to the tangent function.

Lesson 4-2 Review

1. Find the values of x that satisfy the equation $\sin x = \cos x$.

Lesson 4-3: Secant, Cosecant, and Cotangent Functions

All of the other trigonometric functions have as their basis the sine and cosine functions. The secant, cosecant, and cotangent functions are the reciprocals of the cosine, sine, and tangent functions, respectively. We define the secant function in terms of the cosine function:

$$\sec\theta = \frac{1}{\cos\theta}$$

The cosecant function is the reciprocal of the sine function:

$$\csc\theta = \frac{1}{\sin\theta}$$

The cotangent function is the reciprocal of the tangent function:

$$\cot\theta = \frac{1}{\tan\theta}$$

Because the tangent function is itself related to the sine and cosine functions, we can also write the cotangent function in terms of the sine and cosine functions:

$$\cot\theta = \frac{1}{\tan\theta} = \frac{1}{\frac{\sin\theta}{\cos\theta}} = \frac{\cos\theta}{\sin\theta}$$

There are many ways to write the cotangent function. Because $\csc\theta = \frac{1}{\sin\theta}$, we can write the cotangent function as:

$$\cot\theta = \frac{\cos\theta}{\sin\theta} = \cos\theta\csc\theta$$

The trigonometric functions are all related to each other through trigonometric identities, which we will discuss in the next lesson.

The domain of the secant function is the set of all real numbers θ such that $\cos\theta \neq 0$. The domain of the cosecant function is the set of all real numbers θ such that $\sin\theta \neq 0$. Finally, the domain of the cotangent function is the set of all real numbers θ such that $\sin\theta \neq 0$.

θ	$\sin\theta$	$\cos\theta$	$\tan\theta$	$\csc\theta$	$\sec\theta$	$\cot\theta$
0	0	1	0	undefined	1	undefined
$\frac{\pi}{6}$	$\frac{1}{2}$	$\frac{\sqrt{3}}{2}$	$\frac{\sqrt{3}}{3}$	2	$\frac{2\sqrt{3}}{3}$	$\sqrt{3}$
$\frac{\pi}{4}$	$\frac{\sqrt{2}}{2}$	$\frac{\sqrt{2}}{2}$	1	$\sqrt{2}$	$\sqrt{2}$	1
$\frac{\pi}{3}$	$\frac{\sqrt{3}}{2}$	$\frac{1}{2}$	$\sqrt{3}$	$\frac{2\sqrt{3}}{3}$	2	$\frac{\sqrt{3}}{3}$
$\frac{\pi}{2}$	1	0	undefined	1	undefined	0

We can evaluate the secant, cosecant, and cotangent of the angles 0, $\frac{\pi}{6}$, $\frac{\pi}{4}$, $\frac{\pi}{3}$, and $\frac{\pi}{2}$, and expand our table. Again, I would recommend that you become familiar with the trigonometric functions evaluated at these special angles.

The graph of the secant function can be determined from the graph of the cosine function. The domain of the secant function is the set of all angles x for which $\cos x \neq 0$ (the odd half-integer multiples of π), and the secant function will have vertical asymptotes at all of the odd half-integer multiples of π. The period of the secant function is the same as the period of the cosine function: 2π.

The graph of the secant function is shown in Figure 4.7. Remember that the cosine function only takes on values between –1 and 1. Because

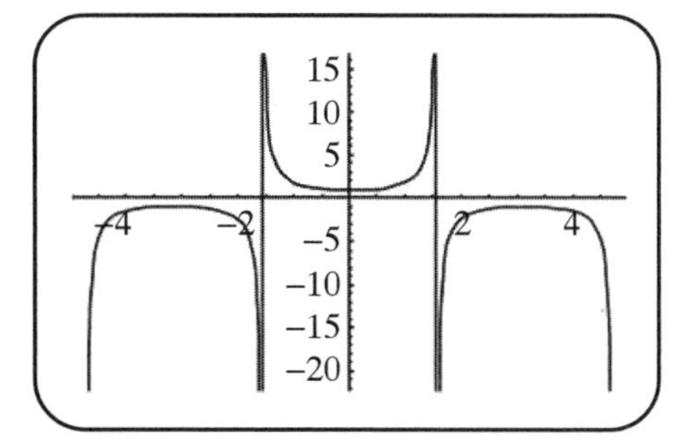

Figure 4.7.

the magnitude of the cosine function is no larger than 1, the magnitude of the secant function will be no smaller than 1. From the definition of the secant function, $\sec x = \frac{1}{\cos x}$, we can see that the graph of the secant function does not have any roots. The secant function inherits the symmetry of the cosine function: the secant function is an even function.

The graph of the cosecant function can be determined from the graph of the sine function. The domain of the cosecant function will be the set of all angles x such that $\sin x \neq 0$. The angles whose sine is 0 are the integer multiples of π. The cosecant function will have vertical asymptotes at all of the integer multiples of π.

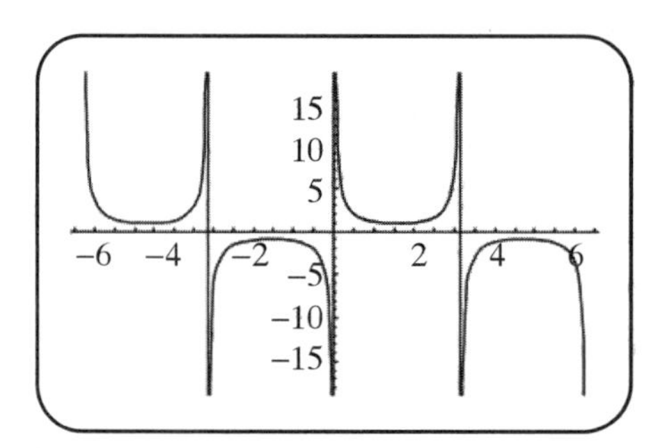

Figure 4.8.

The graph of the cosecant function is shown in Figure 4.8. Remember that the sine function only takes on values between −1 and 1. Because the magnitude of the sine function is no larger than 1, the magnitude of the cosecant function will be no smaller than 1. From the definition of the cosecant function, $\csc x = \frac{1}{\sin x}$, we can see that the graph of the cosecant function does not have any roots. The period of the cosecant function will be the same as the period of the sine function: 2π. The cosecant function inherits the symmetry of the sine function: the cosecant function is an odd function.

The last trigonometric function to graph is the cotangent function. The cotangent function is defined as:

$$\cot x = \frac{1}{\tan x} = \frac{\cos x}{\sin x}$$

Analyzing the cotangent function is very similar to analyzing the tangent function. Some properties are easier to observe by focusing on the cotangent as the reciprocal of the tangent function, and others are easier to observe by writing the cotangent function as a ratio of the cosine function over the sine function.

The domain of the cotangent function will be all values of x for which $\sin x \neq 0$. The cotangent function will have vertical asymptotes at all of the integer multiples of π and zeros at all odd half-integer multiples of π. The cotangent function is an odd function:

$$\cot(-x) = \frac{\cos(-x)}{\sin(-x)} = \frac{\cos x}{-\sin x} = -\cot x$$

The sign of the cotangent function will be the same as the sign of the tangent function: It is positive in quadrants I and III, and negative in quadrants II and IV. The graph of the cotangent function is shown in Figure 4.9.

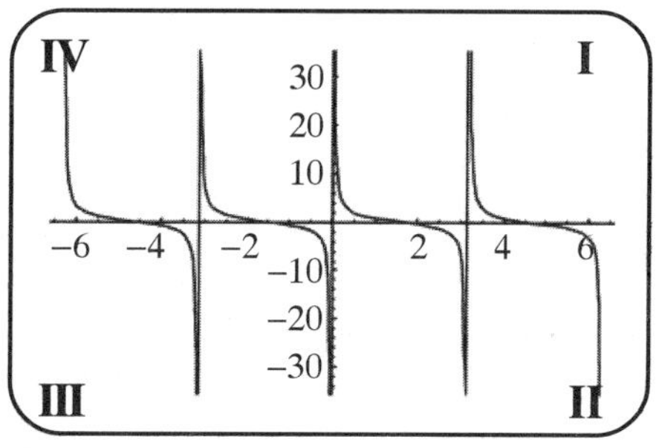

Figure 4.9.

The period of the cotangent function is π (the same as the period of the tangent function). The range of the cotangent function is the set of all real numbers. The cotangent function is periodic, but it does not oscillate the way the sine and cosine functions do.

Lesson 4-4: Identities and Formulas

One of the interesting and useful aspects of the trigonometric functions is their periodic nature, and this periodic behavior can make solving trigonometric functions challenging. Trigonometric identities can help us simplify equations and find their solutions more easily.

There are many relationships between the trigonometric functions. For example, the sine and cosine functions are related to each other through the equation $\sin x = \cos\left(\frac{\pi}{2} - x\right)$. The relationships between trigonometric functions can be written in the form of a trigonometric identity. I have included the most useful identities in this lesson for reference.

A **trigonometric identity** is an equation that is satisfied for all values of the variable. The equation $(x + 1)^2 = x^2 + 2x + 1$ is an example of an identity. This equation is true for all values of x. Equations that are satisfied for particular values of the variable are called **conditional** equations. The equation $x + 1 = 0$ is an example of a conditional equation, because the equation is only satisfied when $x = -1$.

One of the most important trigonometric identities originates from the unit circle definition of the sine and cosine functions:

$$\cos^2\theta + \sin^2\theta = 1$$

Dividing both sides of this equation by $\cos^2\theta$ gives our second identity:

$$\cos^2\theta + \sin^2\theta = 1$$

$$\frac{\cos^2\theta}{\cos^2\theta} + \frac{\sin^2\theta}{\cos^2\theta} = \frac{1}{\cos^2\theta}$$

$$1 + \tan^2\theta = \sec^2\theta$$

This second identity will be surprisingly helpful, and is easy to derive from the first identity.

There are several trigonometric formulas that are important in solving trigonometric equations that you may encounter in calculus. There are sum and difference formulas, double angle formulas, and half-angle formulas.

The sum and difference formulas for the trigonometric functions are summarized in the table that follows:

$$\cos(\alpha+\beta) = \cos\alpha\cos\beta - \sin\alpha\sin\beta \qquad \cos(\alpha-\beta) = \cos\alpha\cos\beta + \sin\alpha\sin\beta$$

$$\sin(\alpha+\beta) = \sin\alpha\cos\beta + \cos\alpha\sin\beta \qquad \sin(\alpha-\beta) = \sin\alpha\cos\beta - \cos\alpha\sin\beta$$

$$\tan(\alpha+\beta) = \frac{\tan\alpha + \tan\beta}{1 - \tan\alpha\tan\beta} \qquad \tan(\alpha-\beta) = \frac{\tan\alpha - \tan\beta}{1 + \tan\alpha\tan\beta}$$

These formulas can be used to find the exact values for the sine of special angles and can be used to prove other useful trigonometric identities.

The double angle formulas for the sine and cosine function can be derived using the sum formulas and letting $\alpha = \beta$:

$$\cos(\alpha+\beta) = \cos\alpha\cos\beta - \sin\alpha\sin\beta \qquad \sin(\alpha+\beta) = \sin\alpha\cos\beta + \cos\alpha\sin\beta$$

$$\cos(\alpha+\alpha) = \cos\alpha\cos\alpha - \sin\alpha\sin\alpha \qquad \sin(\alpha+\alpha) = \sin\alpha\cos\alpha + \cos\alpha\sin\alpha$$

$$\cos(2\alpha) = \cos^2\alpha - \sin^2\alpha \qquad \sin(2\alpha) = 2\sin\alpha\cos\alpha$$

By combining the identity $\cos^2\theta + \sin^2\theta = 1$ with our double angle formula for the cosine function, we can write the double angle in several ways:

$$\cos(2\alpha) = \cos^2\alpha - \sin^2\alpha$$

$$\cos(2\alpha) = 2\cos^2\alpha - 1$$

$$\cos(2\alpha) = 1 - 2\sin^2\alpha$$

The tangent function also has a double angle formula, derived in the same way as the sine and cosine double angle formulas:

$$\tan(2\alpha) = \frac{2\tan\alpha}{1-\tan^2\alpha}$$

The different formulas for the double angle formula for the cosine function are used to derive the half-angle formulas for the sine and cosine functions:

$$\cos\frac{\theta}{2} = \pm\sqrt{\frac{1+\cos\theta}{2}} \qquad \sin\frac{\theta}{2} = \pm\sqrt{\frac{1-\cos\theta}{2}} \qquad \tan\frac{\theta}{2} = \pm\sqrt{\frac{1-\cos\theta}{1+\cos\theta}}$$

There are also product-to-sum and sum-to-product formulas that I will include for reference. These equations are often useful in solving trigonometric expressions.

Product-to-Sum Formulas

$$\sin\alpha\sin\beta = \frac{\cos(\alpha-\beta)-\cos(\alpha+\beta)}{2}$$

$$\cos\alpha\cos\beta = \frac{\cos(\alpha-\beta)+\cos(\alpha+\beta)}{2}$$

$$\sin\alpha\cos\beta = \frac{\sin(\alpha-\beta)+\sin(\alpha+\beta)}{2}$$

Sum-to-Product Formulas

$$\sin\alpha + \sin\beta = 2\sin\left(\frac{\alpha+\beta}{2}\right)\cos\left(\frac{\alpha-\beta}{2}\right)$$

$$\sin\alpha - \sin\beta = 2\sin\left(\frac{\alpha-\beta}{2}\right)\cos\left(\frac{\alpha+\beta}{2}\right)$$

$$\cos\alpha + \cos\beta = 2\cos\left(\frac{\alpha+\beta}{2}\right)\cos\left(\frac{\alpha-\beta}{2}\right)$$

$$\cos\alpha - \cos\beta = -2\sin\left(\frac{\alpha+\beta}{2}\right)\sin\left(\frac{\alpha-\beta}{2}\right)$$

Lesson 4-5: Inverse Trigonometric Functions

Periodic functions can have specific regions where they are increasing or decreasing, but by their very nature, periodic functions cannot be monotonic on their entire domain. A periodic function with period p must satisfy the relationship $f(x+p) = f(x)$. From this we see that $x < x+p$, yet $f(x)$ is not less than $f(x+p)$, so a periodic function cannot be an increasing function on its domain. Similarly, a periodic function cannot be a decreasing function on its domain. Looking at it from another perspective, a periodic function has to repeat, so if a periodic function initially increases, it will have to go down in value somewhere along the line in order for the function to return to its original value.

The periodic nature of the trigonometric functions means that the trigonometric functions are not one-to-one. Functions that are not one-to-one cannot have an inverse, but we can restrict the domain of the trigonometric functions so that they are one-to-one and hence invertible on their restricted domain.

With the function $f(\theta) = \sin\theta$, θ is the measure of an angle, and $f(\theta)$ is a ratio of two lengths (using the triangle definition of the sine function), and is therefore dimensionless. Also, the values that $f(\theta)$ take on must lie between -1 and 1. If we restrict the domain of f so that it is a one-to-one function, then the expression f^{-1} will be a function whose independent variable (or domain) consists of real numbers between -1 and 1, and whose dependent variable (or range) consists of angle measures. We write the inverse sine function as either $\sin^{-1} x$ or $\arcsin x$. The restricted domain for the $f(x) = \sin x$, is $-\frac{\pi}{2} \le x \le \frac{\pi}{2}$, the range is $-1 \le y \le 1$, and the arcsine function is the inverse of the sine function over this restricted domain. The domain of the arcsine function is $-1 \le x \le 1$ and the range is $-\frac{\pi}{2} \le y \le \frac{\pi}{2}$. The equation $y = \arcsin x$ means $x = \sin y$. The graph of the arcsine function is shown in Figure 4.10.

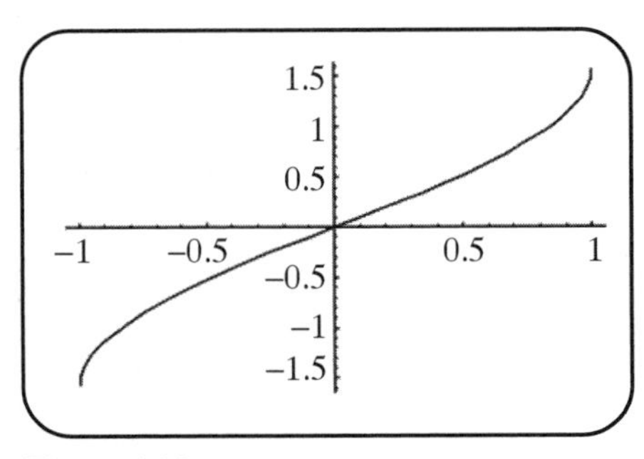

Figure 4.10.

The cosine and the tangent functions also have inverses when their domains are restricted. The arccosine function is the inverse of the cosine function. We write the inverse cosine function as either $\cos^{-1} x$ or $\arccos x$. The restricted domain of $f(x) = \cos x$ is $0 \le x \le \pi$, and the range is $-1 \le y \le 1$. The arccosine function is the inverse of the cosine function on this restricted domain. The domain of the arccosine function is $-1 \leq x \leq 1$, and the range is $0 \leq y \leq \pi$. The restricted domain for the cosine function is different than the restricted domain for the sine function. The equation $y = \arccos x$ means $x = \cos y$. The graph of the arccosine function is shown in Figure 4.11.

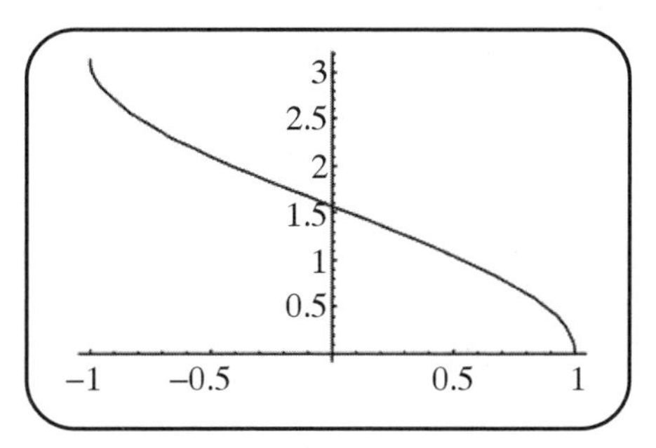

Figure 4.11.

The arctangent function is the inverse of the tangent function. We write the inverse tangent function as either $\tan^{-1} x$ or $\arctan x$. The restricted domain for the arctangent function is $-\frac{\pi}{2} \leq x \leq \frac{\pi}{2}$, and the range of the arctangent function is $(-\infty, \infty)$. The arctangent function is the inverse of the tangent function over this restricted domain. The domain of the arctangent function is $(-\infty, \infty)$, and the range is $-\frac{\pi}{2} \leq y \leq \frac{\pi}{2}$. Notice that the restricted domain for the tangent function is the same as the restricted domain for the sine function. The equation $y = \arctan x$ means $x = \tan y$. The graph of the arctangent function is shown in Figure 4.12.

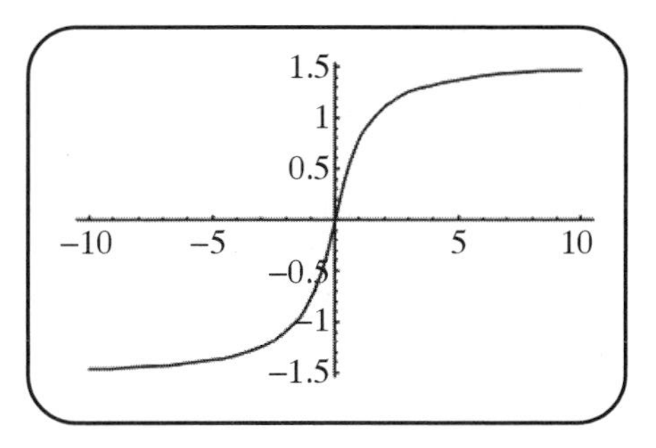

Figure 4.12.

The key to evaluating the inverse trigonometric functions is to be familiar with the trigonometric functions themselves. Knowing the values of the trigonometric functions for the special angles given in the table in Lesson 4-3 is a great place to start!

Answer Key

Lesson 4-2 Review

1. To solve the equation $\sin x = \cos x$, divide both sides of the equation by $\cos x$: $\frac{\sin x}{\cos x} = \frac{\cos x}{\cos x}$. This is equivalent to solving the equation $\tan x = 1$, and from the table in Lesson 4-2, we see that $\tan \frac{\pi}{4} = 1$. The period of the tangent function is π, so the values of x that satisfy the equation $\sin x = \cos x$ are $\frac{\pi}{4} + n\pi$.

Limits

Evaluating a function at a point, or sketching the graph of a function using algebraic techniques, emphasizes the static nature of a function. Calculus takes a more dynamic approach in analyzing functions. Techniques in calculus incorporate the *value* of a function at a point *and* how the function is changing *near that point*. Algebra emphasizes the final destination, and the focus of calculus is the entire trip. As a result, calculus leads us to a more thorough understanding of the properties of functions.

Calculus is the study of limits. In simple terms, a limit allows us to look at what happens to the dependent variable as the independent variable gets close to something. I mentioned earlier that the number e has its basis in calculus. We pushed the formula for compound interest,

$$A = P\left(1+\tfrac{r}{n}\right)^{nt},$$

to an extreme, or to its limit, by compounding interest continuously. As a result, the formula for compound interest transformed into:

$$A = Pe^{rt}.$$

Calculus is all about changing the value of a variable and analyzing what happens as a result. When an object moves, it undergoes a change in position, and calculus can be used to describe this change in location. Calculus is the language of change. Words or phrases such as *approaches*, *gets close to*, and *is near* are all used to represent motion, action, or change. In physics, Newton's laws of motion can be described using calculus. In reaction kinetics, the rate equations are written using calculus. Calculus can be used to describe almost every kind of change that occurs in nature.

Lesson 5-1: Evaluating Limits Numerically

We usually take the limit of a function, or an algebraic expression, as the independent variable heads towards some fixed value. As the independent variable heads towards some fixed value, the independent variable is *changing*, and calculus is used to describe, or quantify, change. Consider the function $f(x) = x + 2$. If x gets close to the number 5, then $f(x)$ gets close to the number 7. In this case, because the rule for the function f is to add 2 to whatever x is, if x is a number close to 5, then $x + 2$ will be a number *close* to, but not equal to, 5 + 2, or 7. We can evaluate the function for various values of x *near*, but not equal to, 5 and see if there is a pattern. We will start by evaluating the function for values of x close to, but always below, 5, as shown in this table.

x	$f(x) = x + 2$
4.9	6.9
4.99	6.99
4.999	6.999

As x approaches 5 from below, meaning that we are letting x take on values that are near, but always less than 5, $f(x)$ gets close to 7. The next table shows the function values when x is close to, but always above, 5.

Once again, as x approaches 5 from above, meaning that we are letting x take on values that are near, but always greater than, 5, $f(x)$ gets close to 7. Because the function gets close to 7 regardless of whether x approaches 5 from below or from above, we say that as x approaches 5, $f(x) = x + 2$ approaches 7. Studying a function's behavior as the independent variable gets close to a fixed number is referred to as finding the limit of the function. Calculus, being the language of change, has its own notation. We write $\lim_{x \to 5}(x+2) = 7$ to represent the idea that as x heads towards 5, the function $f(x) = x + 2$ heads towards 7. We say, "the limit, as x approaches 5, of $x + 2$ is 7."

x	$f(x) = x + 2$
5.1	7.1
5.01	7.01
5.001	7.001

In general, when we write $\lim_{x \to a} f(x) = L$, we mean that as x approaches a, the function $f(x)$ gets close to the number L. We can also say that the limit, as x approaches a, of $f(x)$ is L. Limits deal with the trend of a function, or *where* the function is *trying* to go. Because x can only approach a, and never actually reach it, $f(x)$ will approach L, and may actually never reach it. In fact, a may not even be in the domain of the function $f(x)$, as we will see in the next example.

Example 1

Evaluate $\lim_{x\to 3}\frac{x^2-9}{x-3}$.

Solution: We need to evaluate the function $f(x)=\frac{x^2-9}{x-3}$ for various values of x near 3 and see if there is a trend. We will consider values of x that are less than 3 on the left, and values of x that are greater than 3 on the right:

x	$f(x)=\frac{x^2-9}{x-3}$	x	$f(x)=\frac{x^2-9}{x-3}$
2.9	5.9	3.1	6.1
2.99	5.99	3.01	6.01
2.999	5.999	3.001	6.001

When x approaches 3 from *below*, the function gets closer and closer to 6, and we write $\lim_{x\to 3^-}\frac{x^2-9}{x-3}=6$. The little negative sign above the 3 in the expression for the limit indicates that the limit is one-sided: In this case, it is a limit from below. When x approaches 3 from *above*, the function also gets closer and closer to 6 and we write $\lim_{x\to 3^+}\frac{x^2-9}{x-3}=6$. The little positive sign above the 3 in the expression for the limit indicates that this limit is also one-sided: It is a limit from above. Because the limit from below and the limit from above are the same value, we can write $\lim_{x\to 3}\frac{x^2-9}{x-3}=6$, leaving off the + or – above the 3.

In the previous example, notice that $x = 3$ is not in the domain of the function $f(x)=\frac{x^2-9}{x-3}$. If we evaluated $f(x)=\frac{x^2-9}{x-3}$ at $x = 3$, we would get $f(3)=\frac{3^2-9}{3-3}=\frac{0}{0}$. Any *non-zero* number divided by itself is 1, but $\frac{0}{0}$ is considered to be an **indeterminate expression.** This is one of the advantages of limits: x may be approaching a number that is not in the domain, but we can still observe how the function behaves *around* that number.

To evaluate limits numerically, set up a table similar to the table used in Example 1. Evaluate the limit from below and the limit from above. If both limits approach the same value, then the limit exists and is equal to that value.

Example 2

Evaluate $\lim_{x \to 0} \frac{2^x - 1}{x}$.

Solution: We need to evaluate the function $f(x) = \frac{2^x - 1}{x}$ for various values of x near 0 and see if there is a trend. We will consider values of x that are less than 0 on the left, and values of x that are greater than 0 on the right:

x	$f(x) = \frac{2^x - 1}{x}$	x	$f(x) = \frac{2^x - 1}{x}$
–0.1	0.66967	0.1	0.717735
–0.01	0.69075	0.01	0.695555
–0.001	0.692907	0.001	0.693387

When x approaches 0 from *below*, the function gets closer and closer to 0.693, so $\lim_{x \to 0^-} \frac{2^x - 1}{x} = 0.693$ (which is a one-sided limit). When x approaches 0 from *above*, the function also gets closer and closer to 0.693, so $\lim_{x \to 0^+} \frac{2^x - 1}{x} = 0.693$ (which is a one-sided limit). Because the limit from below and the limit from above are the same value, we can write $\lim_{x \to 0} \frac{2^x - 1}{x} = 0.693$.

The limit in Example 1 was a whole number (6), but the limit in Example 2 was not (0.693). The function values in Example 2 were approaching a number that you may not be familiar with. If a more accurate value for the limit is needed, evaluate the function for values of x that are even closer to 0, such as $x = \pm 0.0000001$.

Example 3

Evaluate $\lim_{x \to 4} \frac{\sqrt{x}-2}{x-4}$.

Solution: We need to evaluate the function $f(x)=\frac{\sqrt{x}-2}{x-4}$ for various values of x near 4 and see if there is a trend. We will consider values of x that are less than 4 on the left, and values of x that are greater than 4 on the right:

x	$f(x)=\frac{\sqrt{x}-2}{x-4}$	x	$f(x)=\frac{\sqrt{x}-2}{x-4}$
3.9	0.25158	4.1	0.24846
3.99	0.25016	4.01	0.24984
3.999	0.25002	4.001	0.24998

When x approaches 4 from below, the function gets closer and closer to 0.25, and we write $\lim_{x \to 4^-} \frac{\sqrt{x}-2}{x-4} = 0.25$. When x approaches 4 from above, the function also gets closer and closer to 0.25, and we write $\lim_{x \to 4^+} \frac{\sqrt{x}-2}{x-4} = 0.25$. Because the limit from below and the limit from above are the same value, we can write $\lim_{x \to 4} \frac{\sqrt{x}-2}{x-4} = 0.25$.

We have evaluated limits for three types of functions: a rational function, one that involves an exponential function, and one that involves radicals. We should also practice working with trigonometric and logarithmic functions.

Example 4

Evaluate $\lim_{x \to 0} \frac{\sin x}{x}$.

Solution: We need to evaluate the function $f(x)=\frac{\sin x}{x}$ for various values of x near 0 and see if there is a trend. We will consider values of x that are less than 0 on the left, and values of x that are greater than 0 on the right:

x	$f(x)=\frac{\sin x}{x}$	x	$f(x)=\frac{\sin x}{x}$
–0.1	0.998334	0.1	0.998334
–0.01	0.999983	0.01	0.999983
–0.001	0.99999983	0.001	0.99999983

Notice the symmetry of the table of values. Because $\sin x$ and x are both odd functions, their ratio will be an even function. An even function is symmetric about the y-axis, with $f(-x)=f(x)$. This is a situation in which recognizing the symmetry of a function will save us work. When x approaches 0 from below, the function gets closer and closer to 1, and we write $\lim_{x\to 0^-}\frac{\sin x}{x}=1$. When x approaches 0 from above, the function also gets closer and closer to 1, and we write $\lim_{x\to 0^+}\frac{\sin x}{x}=1$. Because the limit from below and the limit from above are the same value, we can write $\lim_{x\to 0}\frac{\sin x}{x}=1$. This is an important limit to be familiar with.

Example 5

Evaluate $\lim_{x\to 0}\frac{\ln(x+1)}{x}$.

Solution: We need to evaluate the function $f(x)=\frac{\ln(x+1)}{x}$ for various values of x near 0 and see if there is a trend. We will consider values of x that are less than 0 on the left, and values of x that are greater than 0 on the right:

x	$f(x)=\frac{\ln(x+1)}{x}$	x	$f(x)=\frac{\ln(x+1)}{x}$
–0.1	1.053605	0.1	0.953102
–0.01	1.005033	0.01	0.995033
–0.001	1.000500	0.001	0.999500

When x approaches 0 from below, the function gets closer and closer to 1, and we write $\lim_{x\to 0^-}\frac{\ln(x+1)}{x}=1$. When x approaches 0 from above, the function also gets closer and closer to 1, and we write $\lim_{x\to 0^+}\frac{\ln(x+1)}{x}=1$. Because the limit from below and the limit from above are the same value, we can write $\lim_{x\to 0}\frac{\ln(x+1)}{x}=1$.

The previous examples all have one thing in common: The value that x approaches is not in the domain of the function. When evaluating $\lim_{x\to 3}\frac{x^2-9}{x-3}$, notice that as x approaches 3, $x^2 - 9$ heads towards 0 and $x - 3$ also heads towards 0. The overall function, however, approaches 6. When evaluating $\lim_{x\to 0}\frac{2^x-1}{x}$, notice that as x approaches 0, $2^x - 1$ also heads towards 0, but the overall function approaches 0.693. When evaluating $\lim_{x\to 4}\frac{\sqrt{x}-2}{x-4}$, as x approaches 4, $\sqrt{x}-2$ heads towards 0 and $x - 4$ also heads towards 0. The overall function approaches 0.25. The same holds true for $\lim_{x\to 0}\frac{\sin x}{x}$ and $\lim_{x\to 0}\frac{\ln(x+1)}{x}$. In all of these examples, the numerator and the denominator of the function both approach 0, but the overall ratio approaches some finite value. If a point is not in the domain of a function, but the limit of the function as x approaches that point is finite, we call that point a **removable singularity**. A **singularity** is a point at which a function is not defined. A removable singularity is a singularity that can be removed, as we will see in Chapter 7. In the event that, as x approaches a point a, the denominator of a function approaches 0 while the numerator approaches a *non-zero* number, then the point a is a **non-removable singularity**, and the function will have a vertical asymptote at $x = a$.

Lesson 5-1 Review

1. Find the following limits numerically:

   a. $\lim_{x \to 2} \frac{4 - x^2}{x - 2}$

   b. $\lim_{x \to 0} \frac{3^x - 1}{x}$

   c. $\lim_{x \to 9} \frac{\sqrt{x} - 3}{x - 9}$

   d. $\lim_{x \to 0} \frac{\cos x - 1}{x}$

   e. $\lim_{x \to 1} \frac{\ln(2x - 1)}{x - 1}$

   f. $\lim_{x \to 0} \frac{e^x - 1}{x}$

2. Is the point $x = 2$ a removable or non-removable singularity of the function $f(x) = \frac{\sqrt{x - 2}}{x - 2}$?

Lesson 5-2: Evaluating Limits Graphically

If a function is defined graphically, evaluating a limit involves tracing the function along its graph. Trace the graph from the left to find the limit from below, and trace it from the right to find the limit from above.

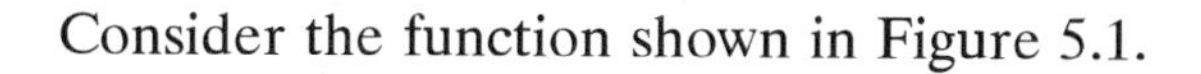

Consider the function shown in Figure 5.1.

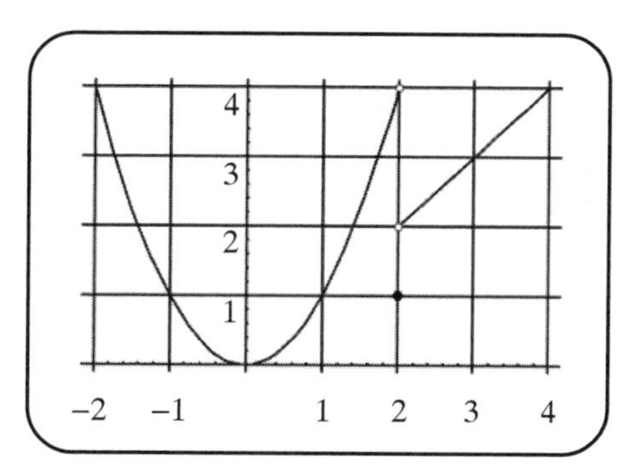

Figure 5.1.

Some interesting things happen with this function at $x = 2$. Notice that as x approaches 2 from below (or from the left), we are on the parabolic part of the graph of the function, which gets closer and closer to 4: $\lim_{x \to 2^-} f(x) = 4$. The function never reaches 4, because as soon as x reaches 2, the function drops abruptly down to 1. That's what the closed circle at the point (2, 1) represents. To evaluate $\lim_{x \to 2^+} f(x)$, we need to trace the function from above (or from the right), choosing values of x that are close to 2, but that are greater than 2. In this region, we must focus on the linear part of the graph of the function, which gets closer and closer to 2 as x heads towards 2: $\lim_{x \to 2^+} f(x) = 2$. Notice that, with this function, the limit from the left and the limit from the right exist but are not the same value. When the limit from below and the limit from above exist but are not equal, we say that $\lim_{x \to 2} f(x)$ does not exist. Also, with this function, neither the limit from the left nor the limit from the right are equal to the function value at $x = 2$. The point $x = 2$ is truly a problematic point for this function.

We can examine how this function behaves at other values of x. For example, as x approaches 1 from below (or from the left), we see that the function gets closer and closer to 1. As x approaches 1 from above (or from the right), we see that the function also gets closer and closer to 1. We can write $\lim_{x \to 1} f(x) = 1$. Similarly, we see that $\lim_{x \to -1} f(x) = 1$. The difference in the behavior of the function at $x = 1$ and $x = 2$ is apparent in the graph: At $x = 2$ the function changes abruptly; the two pieces of the function do not join together at $x = 2$. As we will learn in Chapter 7, the point $x = 2$ is a discontinuity of the function.

Example 2

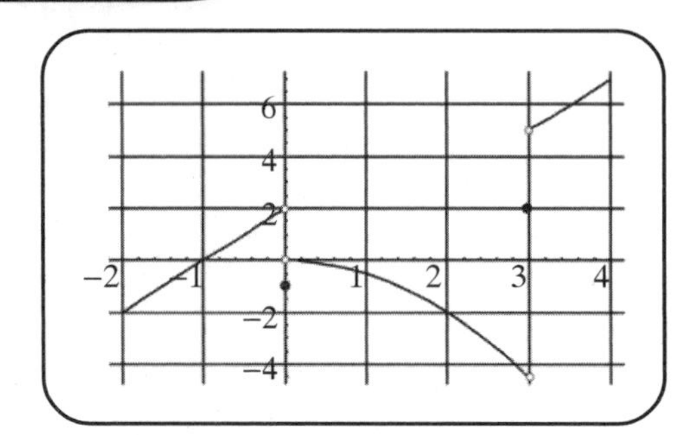

Figure 5.2.

For the function shown in Figure 5.2, evaluate the following:

a. $\lim_{x\to 0^-} f(x)$, $\lim_{x\to 0^+} f(x)$, $\lim_{x\to 0} f(x)$, and $f(0)$

b. $\lim_{x\to 2^-} f(x)$, $\lim_{x\to 2^+} f(x)$, $\lim_{x\to 2} f(x)$, and $f(2)$

c. $\lim_{x\to 3^-} f(x)$, $\lim_{x\to 3^+} f(x)$, $\lim_{x\to 3} f(x)$, and $f(3)$

Solution:

a. $\lim_{x\to 0^-} f(x) = 2$, $\lim_{x\to 0^+} f(x) = 0$, $\lim_{x\to 0} f(x)$ does not exist, because the limit from below and the limit from above are not equal to each other, and $f(0) = -1$.

b. $\lim_{x\to 2^-} f(x) = -2$, $\lim_{x\to 2^+} f(x) = -2$, $\lim_{x\to 2} f(x) = -2$, and $f(2) = -2$

c. $\lim_{x\to 3^-} f(x) = -4.5$, $\lim_{x\to 3^+} f(x) = 5$, $\lim_{x\to 3} f(x)$ does not exist, because the limit from below and the limit from above are not equal to each other, and $f(3) = 2$.

Lesson 5-3: Evaluating Limits Algebraically

There are several methods that can be used to find the limit of a function that is defined algebraically, or by a formula. We can examine some special functions and make some observations.

Let's begin by exploring the limit of a constant function. For example, $\lim_{x\to 2} 5 = 5$. In other words, as x gets close to 2, the function $f(x) = 5$ gets close to 5. The constant function does not care what value x gets close to. The constant function is just that: constant. Regardless of what value x gets close to, a constant function stays where it is. In general, if a and c are any constants, then $\lim_{x\to a} c = c$. That is one of our first rules for evaluating a limit algebraically.

The next function to examine is the function $f(x) = x$ as x gets close to 4: As x gets close to 4, the function $f(x) = x$ gets close to 4. In other words, $\lim_{x\to 4} x = 4$. In general, if a is any constant, then $\lim_{x\to a} x = a$.

We can now examine the function $f(x) = x^2$ as x gets close to -2: As x gets close to -2, $f(x) = x^2$ gets close to $(-2)^2$, or 4. We can write $\lim_{x\to -2} x^2 = (-2)^2 = 4$. In general, $\lim_{x\to a} x^2 = a^2$.

For simple monomials, $\lim_{x\to a} x^n = a^n$. This is an important observation, and it leads us to our first technique: substituting the value that x approaches into the formula for the function to determine the value that the function approaches: $\lim_{x\to a} f(x) = f(a)$.

If this technique worked all of the time, we would not need any other methods. Unfortunately, there are times when this approach fails. We saw examples of this when we evaluated $\lim_{x\to 3} \frac{x^2-9}{x-3}$, $\lim_{x\to 0} \frac{2^x-1}{x}$, and $\lim_{x\to 4} \frac{\sqrt{x}-2}{x-4}$. In those examples, the value that x was approaching was a singularity, or a point where the function was not defined. In those examples, both the numerator and the denominator of the function headed towards 0, and we had to use a numerical technique to determine the limit. There are algebraic techniques to evaluate two of these three limits, and evaluating the third will require us to examine a well-known transcendental number in more detail.

To simplify the process of evaluating limits algebraically, we can make use of some established properties of limits. Suppose f and g are two functions, c and k are constants, and $\lim_{x\to c} f(x) = L$ and $\lim_{x\to c} g(x) = M$, where L and M are real numbers. Then the following statements are true:

1. $\lim_{x\to c}\left[f(x) \pm g(x)\right] = \lim_{x\to c} f(x) \pm \lim_{x\to c} g(x) = L \pm M$: The limit of the sum (or difference) of two functions is the sum (or difference) of the limits of the functions.

2. $\lim_{x\to c}\left[k \cdot f(x)\right] = k \cdot \lim_{x\to c} f(x) = k \cdot L$: The limit of the product of a constant and a function is the product of the constant and the limit of the function.

3. $\lim_{x\to c}\left[f(x) \cdot g(x)\right] = \lim_{x\to c} f(x) \cdot \lim_{x\to c} g(x) = L \cdot M$: The limit of the product of two functions is the product of the limits of the two functions.

4. $\lim_{x\to c} \sqrt[n]{f(x)} = \sqrt[n]{L}$: The limit of the root of a function is the root of the limit of the function. Here we require L to be positive if n is even.

5. $\lim_{x\to c} \frac{f(x)}{g(x)} = \frac{\lim_{x\to c} f(x)}{\lim_{x\to c} g(x)} = \frac{L}{M}$: The limit of the ratio of two functions is the ratio of the limits of the two functions. Here we require that $M \neq 0$.

Example 1

Use the properties of limits to evaluate $\lim_{x\to 3}\left(x^2 - 2x\right)$.

Solution:

Property 1 $\qquad \lim_{x\to 3}\left(x^2 - 2x\right) = \lim_{x\to 3} x^2 - \lim_{x\to 3} 2x$

Properties 2 and 3 $\qquad \lim_{x\to 3} x^2 - \lim_{x\to 3} 2x = \left(\lim_{x\to 3} x\right) \cdot \left(\lim_{x\to 3} x\right) - 2 \cdot \lim_{x\to 3} x$

Using the fact that $\lim_{x \to a} x = a$ $\left(\lim_{x \to 3} x\right) \cdot \left(\lim_{x \to 3} x\right) - 2 \cdot \lim_{x \to 3} x = 3 \cdot 3 - 2 \cdot 3 = 3$

From this we see that $\lim_{x \to 3}\left(x^2 - 2x\right) = 3$.

Evaluating $\lim_{x \to 3}\left(x^2 - 2x\right)$ can also be done using a direct substitution technique: Substitute $x = 3$ into the function $f(x) = x^2 - 2x$ directly: $f(3) = 3^2 - 2 \cdot 3 = 9 - 6 = 3$. This direct substitution method is the one to try first. If the result is a number, and not an indeterminate expression, your work is done. If the result is an expression of the form $\frac{0}{0}$, then this technique will not work and you will need to try another approach.

For example, if we wanted to evaluate $\lim_{x \to 3} \frac{x^2 - 9}{x - 3}$, we would first try the direct substitution method: Evaluate the function $\frac{x^2 - 9}{x - 3}$ at $x = 3$. In this case, however, the result is $\frac{0}{0}$, so we need another approach. We have already evaluated this limit numerically, but now we would like to evaluate it using a different technique. We can use the properties of limits, along with some algebraic simplification, to evaluate $\lim_{x \to 3} \frac{x^2 - 9}{x - 3}$ algebraically. Notice that the numerator of the function $\frac{x^2 - 9}{x - 3}$ is the difference between two squares. If we factored the numerator, we would have $\frac{(x-3)(x+3)}{(x-3)}$. At this point, we are tempted to cancel the factor $(x - 3)$ that appears in both the numerator and the denominator, and, as long as $x \neq 3$, we can. Recall that $\lim_{x \to 3} \frac{x^2 - 9}{x - 3}$ means that x gets close to, but never actually gets to reach, the number 3, and because of this we are allowed to cancel the common factor and try to evaluate the limit again using direct substitution:

$$\lim_{x \to 3} \frac{x^2 - 9}{x - 3} = \lim_{x \to 3} \frac{\cancel{(x-3)}(x+3)}{\cancel{(x-3)}} = \lim_{x \to 3}(x + 3) = 6$$

This answer agrees with what we observed previously. This method of evaluating limits makes use of *factoring*.

Example 2

Use the properties of limits to evaluate $\lim_{x \to 1} \frac{x-1}{x^2+x-2}$.

Solution: First, try the method of direct substitution:

Evaluate $f(x) = \frac{x-1}{x^2+x-2}$ at $x = 1$.

Because the result is an indeterminate of the form $\frac{0}{0}$, this technique will not work.

The denominator of this function is a quadratic expression that can be factored:

$$f(x) = \frac{x-1}{x^2+x-2} = \frac{(x-1)}{(x-1)(x+2)}$$

and both the numerator and the denominator have a common factor of $(x - 1)$ that can be cancelled if $x \neq 1$. This will enable us to evaluate the limit:

$$\lim_{x \to 1} \frac{x-1}{x^2+x-2} = \lim_{x \to 1} \frac{\cancel{(x-1)}}{\cancel{(x-1)}(x+2)} \lim_{x \to 1} \frac{1}{x+2} = \frac{1}{3}$$

Factoring is not the only algebraic simplification that we can try. Let's evaluate $\lim_{x \to 4} \frac{\sqrt{x}-2}{x-4}$. First, try the direct substitution method. Because both the numerator and the denominator of the function $f(x) = \frac{\sqrt{x}-2}{x-4}$ head towards 0 as x approaches 4, the direct substitution method will not work. The function $f(x) = \frac{\sqrt{x}-2}{x-4}$ cannot be factored by our traditional methods, so we will need another technique. This next technique is only useful when evaluating limits of functions that involve radicals, and it involves rationalizing the numerator.

The **conjugate** of a binomial expression $(a + b)$ is $(a - b)$. The conjugate of $(\sqrt{x} - 2)$ is $(\sqrt{x} + 2)$. We can multiply both the numerator and the denominator of the function $f(x)=\frac{\sqrt{x}-2}{x-4}$ by the conjugate of the numerator:

$$f(x)=\frac{\sqrt{x}-2}{x-4}\cdot\frac{\sqrt{x}+2}{\sqrt{x}+2}.$$

It may look as though we have made things worse, but things will get better once we multiply the two numerators together:

$$f(x)=\frac{\sqrt{x}-2}{x-4}\cdot\frac{\sqrt{x}+2}{\sqrt{x}+2}=\frac{(x-4)}{(x-4)(\sqrt{x}+2)}$$

Notice that I kept the denominator factored. The reason is that now both the numerator and the denominator have a common factor: $(x - 4)$. This common factor can be cancelled, as long as $x \neq 4$. Because we are taking the limit as x approaches (but never reaches) 4, we can cancel out the common factor and evaluate the limit directly:

$$\lim_{x\to 4}\frac{\sqrt{x}-2}{x-4}=\lim_{x\to 4}\frac{\cancel{(x-4)}}{\cancel{(x-4)}(\sqrt{x}+2)}=\lim_{x\to 4}\frac{1}{\sqrt{x}+2}=\frac{1}{\sqrt{4}+2}=\frac{1}{4}$$

This was the same result that we obtained previously.

Example 3

Use the properties of limits to evaluate $\lim_{x\to 0}\frac{\sqrt{x+9}-3}{x}$.

Solution: First, try to use the substitution method. In this case, both the numerator and the denominator head towards 0 as x approaches 0, so the substitution method will not work. The function $f(x)=\frac{\sqrt{x+9}-3}{x}$ is not a rational function, and trying to factor it will not get you very far. Because the first two techniques have failed us, and because there are roots involved in this limit, we will reach for our conjugate method. Multiply the numerator and the denominator by the conjugate of $\sqrt{x+9}-3$, which is $\sqrt{x+9}+3$, and simplify:

Multiply both the numerator and the denominator by $\sqrt{x+9}+3$

$$\lim_{x\to 0}\frac{\sqrt{x+9}-3}{x}=\lim_{x\to 0}\frac{\sqrt{x+9}-3}{x}\cdot\frac{\sqrt{x+9}+3}{\sqrt{x+9}+3}$$

Expand the numerator and keep the denominator factored

$$\lim_{x\to 0}\frac{(x+9)-9}{x\left(\sqrt{x+9}+3\right)}$$

Cancel the common factor x

$$\lim_{x\to 0}\frac{\not{x}}{\not{x}\left(\sqrt{x+9}+3\right)}$$

Evaluate the limit using substitution

$$\lim_{x\to 0}\frac{1}{\sqrt{x+9}+3}=\frac{1}{\sqrt{0+9}+3}=\frac{1}{6}$$

Thus $\lim_{x\to 0}\frac{\sqrt{x+9}-3}{x}=\frac{1}{6}$.

We now have three methods for evaluating limits algebraically: direct substitution, factoring, and multiplication by the conjugate. The substitution method applies if the denominator of the function does not approach 0. The factoring method works for rational functions, and the conjugate method works when either the numerator or the denominator involves a radical. If all three of these methods fail, we can try to find the limit numerically, as we did at the beginning of this chapter.

Lesson 5-3 Review

Evaluate the following limits:

1. $\lim_{x\to 2}\frac{x+1}{\sqrt{x^2+5}-2}$

2. $\lim_{x\to 3}\frac{x-3}{x^2-x-6}$

3. $\lim_{x\to 0}\frac{\sqrt{2x+1}-1}{x}$

4. $\lim_{x\to -3}\frac{x+3}{\sqrt{x+7}-2}$

Lesson 5-4: Evaluating Limits That Involve Infinity

Limits are all about what a function or a variable approaches, yet never reaches. Infinity is a similar concept. Functions or variables can never *equal* infinity, but they can approach it by getting very, very large. We have already touched on these ideas when we studied the asymptotic behavior of a function.

Vertical asymptotes come about when a function heads towards $\pm\infty$ as x approaches a fixed number a, either from below or from above. In other words, a function will have a vertical asymptote if $\lim_{x\to a^-} f(x) = \pm\infty$ or $\lim_{x\to a^+} f(x) = \pm\infty$. We have already discussed how to find the vertical asymptotes of a rational function, and the same ideas apply for finding vertical asymptotes of all functions. Look for places where the denominator of the function approaches 0 while the numerator of the function does *not* approach 0. If possible, factor the function and look for the zeros of the denominator. Evaluate the limit of the function as x approaches each of the zeros of the denominator. If the limit exists and is a finite value, the point will be a removable singularity. If the limit does not exist because the function heads towards $\pm\infty$, then the point is a vertical asymptote.

Finding the horizontal asymptotes of a function involves analyzing that function for values of x that are large in magnitude. In other words, finding the horizontal asymptotes of a function involves evaluating $\lim_{x\to\infty} f(x)$ or $\lim_{x\to-\infty} f(x)$. If this limit exists and is a finite number, then the function will have a horizontal asymptote. Use any of the algebraic techniques you have learned to simplify the function and evaluate the limit. You may need to factor, rationalize the numerator or the denominator, or use your understanding of the asymptotic behavior of power functions or polynomials in the process. We have already established that $\lim_{x\to\infty} x^p = 0$ if $p < 0$, which is equivalent to $\lim_{x\to\infty} \frac{1}{x^p} = 0$ if $p > 0$.

Example 1

Evaluate $\lim_{x\to\infty} \sqrt{x+2} - \sqrt{x}$.

Solution: Rationalize the numerator and use the asymptotic properties of power functions:

Multiply by the conjugate of $\sqrt{x+2}-\sqrt{x}$

$$\lim_{x\to\infty}\left(\sqrt{x+2}-\sqrt{x}\right)\cdot\frac{\sqrt{x+2}+\sqrt{x}}{\sqrt{x+2}+\sqrt{x}}$$

Multiply the numerators together and keep the denominator factored

$$\lim_{x\to\infty}\frac{(x+2)-x}{\left(\sqrt{x+2}+\sqrt{x}\right)}$$

Simplify the numerator
$$\lim_{x\to\infty}\frac{2}{\left(\sqrt{x+2}+\sqrt{x}\right)}$$

Use the asymptotic properties of polynomials:

As x gets large, $\sqrt{x+2}\sim\sqrt{x}$, so $\frac{2}{\left(\sqrt{x+2}+\sqrt{x}\right)}\sim\frac{2}{2\sqrt{x}}\to 0$ as $x\to\infty$.

Therefore: $\lim_{x\to\infty}\sqrt{x+2}-\sqrt{x}=\lim_{x\to\infty}\frac{2}{\left(\sqrt{x+2}+\sqrt{x}\right)}=0$.

Answer Key

Lesson 5-1 Review

1. a. $\lim_{x\to 2}\frac{4-x^2}{x-2}=-4$

   b. $\lim_{x\to 0}\frac{3^x-1}{x}\approx 1.099$

   c. $\lim_{x\to 9}\frac{\sqrt{x}-3}{x-9}\approx 0.1667$

   d. $\lim_{x\to 0}\frac{\cos x-1}{x}=0$

   e. $\lim_{x\to 1}\frac{\ln(2x-1)}{x-1}=2$

   f. $\lim_{x\to 0}\frac{e^x-1}{x}=1$

2. The point $x = 2$ is a non-removable singularity: $\lim_{x\to 2}\frac{\sqrt{x-2}}{x-2} = \lim_{x\to 2}\frac{1}{\sqrt{x-2}}$, and this limit does not exist.

Lesson 5-3 Review

1. Use substitution: $\lim_{x\to 2}\frac{x+1}{\sqrt{x^2+5}-2} = \frac{2+1}{\sqrt{2^2+5}-2} = 3$

2. Factor and cancel: $\lim_{x\to 3}\frac{x-3}{x^2-x-6} = \lim_{x\to 3}\frac{\cancel{(x-3)}}{\cancel{(x-3)}(x+2)} = \lim_{x\to 3}\frac{1}{(x+2)} = \frac{1}{5}$

3. Multiply by the conjugate of the numerator:

$$\lim_{x\to 0}\frac{\sqrt{2x+1}-1}{x} = \lim_{x\to 0}\left(\frac{\sqrt{2x+1}-1}{x}\cdot\frac{\sqrt{2x+1}+1}{\sqrt{2x+1}+1}\right) = \lim_{x\to 0}\left(\frac{(2x+1)-1}{x\left(\sqrt{2x+1}+1\right)}\right)$$

$$= \lim_{x\to 0}\left(\frac{2\cancel{x}}{\cancel{x}\left(\sqrt{2x+1}+1\right)}\right) = \frac{2}{2} = 1$$

4. Multiply by the conjugate of the denominator:

$$\lim_{x\to -3}\frac{x+3}{\sqrt{x+7}-2} = \lim_{x\to -3}\left(\frac{x+3}{\sqrt{x+7}-2}\cdot\frac{\sqrt{x+7}+2}{\sqrt{x+7}+2}\right) = \lim_{x\to -3}\left(\frac{(x+3)\sqrt{x+7}+2}{(x+7)-4}\right) = \lim_{x\to -3}\left(\frac{\cancel{(x+3)}\sqrt{x+7}+2}{\cancel{(x+3)}}\right) = 4$$

Continuity

We can use the notion of a limit to develop the idea of continuity. It is possible for the graph of a function to have breaks, holes, and jumps. The limit of a function does not exist at the breaks and jumps, because the limit from below and the limit from above are not equal. Holes are a problem because they represent points that are either not in the domain of the function or where the function changes at an instant. Breaks and holes are called points of discontinuity. The discontinuities of a function are the points where the function has a break or a hole in its graph. Functions are usually well-behaved away from the breaks, jumps, and holes, but we need a mathematical way to describe what this means.

Lesson 6-1: Continuity and Limits

Conceptually, a continuous function is a function that has no breaks, holes, or jumps. Its graph can be drawn completely without lifting the pencil off of the page. This conceptual idea of continuity needs to be described more precisely, and limits play a crucial role in defining continuity.

In order for a function to be continuous at a point, we need the limit from the left and the limit from the right to be the same, so that the limit exists. This requirement avoids the situation that arises when two parts of our function do not join together. We also need the function to be defined at the point. This requirement takes care of discontinuities that result from unfilled holes and vertical asymptotes. The third requirement is that the limit and the value of the function must be the same. This third requirement fills in any holes. These three requirements can be stated using limits.

A function $f(x)$ is **continuous** at a point $x = a$ if three conditions hold:

- $\lim_{x \to a} f(x)$ exists
- $f(a)$ is defined
- $\lim_{x \to a} f(x) = f(a)$

If one or more of the three conditions in the definition does not hold, the function is **discontinuous** at $x = a$.

Continuity is an important property of a function. The third requirement for continuity is what allows us to evaluate limits using the substitution method, as we discussed in Chapter 5. We can determine whether the *graph* of a function is continuous simply by inspection. The function will be discontinuous wherever the graph of the function has any breaks, holes, jumps, or vertical asymptotes. We will discuss the continuity of some important elementary functions and discover the best approach to finding limits of these functions.

Let's start with polynomials. Polynomials are continuous on their domain. This means that if you ever need to evaluate the limit of a polynomial, the substitution method will always work.

Rational functions, which involve the ratio of two polynomials, will be continuous everywhere except where the denominator is equal to 0. To evaluate the limit of a rational function, check to see if the limit of the polynomial in the denominator goes to 0: If it does not, use the substitution method. If it does, check to see if the numerator also goes to 0: If it does not, then the limit will not exist. If both the numerator and the denominator head towards 0, then try factoring both polynomials. Remember that $(x - a)$ is a factor of a polynomial $P(x)$ if and only if $P(a) = 0$. If both the numerator and the denominator head towards 0 as x approaches a, then $(x - a)$ will be a factor of both the numerator and the denominator. Divide each of the polynomials by $(x - a)$ and try to evaluate the limit of the resulting rational function. This process is illustrated in Example 1.

Example 1

Evaluate $\lim_{x \to 2} \frac{x^3 - 3x^2 + 4}{x^3 - 2x^2 - 4x + 8}$.

Solution: Try the substitution method first. Evaluate the limit of the numerator and the denominator:

$$\lim_{x\to 2} x^3 - 3x^2 + 4 = (2)^3 - 3(2)^2 + 4 = 8 - 12 + 4 = 0$$

$$\lim_{x\to 2} x^3 - 2x^2 - 4x + 8 = (2)^3 - 2(2)^2 - 4(2) + 8 = 8 - 8 - 8 + 8 = 0$$

Both the numerator and the denominator head towards 0 as x approaches 2, so the polynomials in the numerator and the denominator have a common factor $(x - 2)$. We will need to divide the numerator and the denominator by $(x - 2)$:

$$\frac{x^3 - 3x^2 + 4}{(x-2)} = x^2 - x - 2$$

$$\frac{x^3 - 2x^2 - 4x + 8}{(x-2)} = x^2 - 4$$

Now we are ready to simplify and try again:

$$\lim_{x\to 2}\frac{x^3 - 3x^2 + 4}{x^3 - 2x^2 - 4x + 8} = \lim_{x\to 2}\frac{(x^2 - x - 2)}{(x^2 - 4)}$$

Try the substitution method again, by evaluating the limit of the new numerator and the new denominator:

$$\lim_{x\to 2} x^2 - x - 2 = (2)^2 - 2 - 2 = 0$$

$$\lim_{x\to 2} x^2 - 4 = (2)^2 - 4 = 0$$

Because both limits equal 0, we need to divide again:

$$\frac{x^2 - x - 2}{(x-2)} = x + 1$$

$$\frac{x^2 - 4}{(x-2)} = x + 2$$

Now we are ready to simplify and try to take the limit again:

$$\lim_{x\to 2}\frac{x^2 - x - 2}{x^2 - 4} = \lim_{x\to 2}\frac{(x+1)}{(x+2)} = \frac{3}{4}$$

We see that $\lim_{x\to 2}\frac{x^3 - 3x^2 + 4}{x^3 - 2x^2 - 4x + 8} = \frac{3}{4}$.

Exponential functions are continuous on their domain and are never equal to 0. So even if an exponential function appears in the denominator of a function, it will not cause any problems. Logarithmic functions, being inverses of exponential functions, are also continuous on their domain. Because $\log_b 1 = 0$, there may be some problems with evaluating limits if a logarithmic function appears in the denominator of a function.

Finally, we will discuss the trigonometric functions. The sine and cosine functions are continuous on their domain, so the substitution method will work when evaluating limits of these functions. The other trigonometric functions involve ratios of the sine and cosine functions, and the substitution method can fail when evaluating limits if the denominator approaches 0.

When evaluating the limit of a function, pay attention to the location of any discontinuities. If x is not approaching a discontinuity of the function, use the substitution method to evaluate the limit. If x is approaching a discontinuity of the function, look for a vertical asymptote. Vertical asymptotes are characterized by the denominator of the function heading toward 0 while the numerator heads towards some non-zero number. If both the numerator and the denominator head towards 0, reach for another method, such as factoring or rationalizing either the numerator or the denominator. If those methods fail (or do not apply), try evaluating the limit numerically by filling out a table similar to those we worked with in Chapter 5.

The last functions to explore are piecewise-defined functions. The most likely place for a piecewise-defined function to be discontinuous is where the two pieces come together. If both pieces of a piecewise-defined function meet up at their junction, then the function will be continuous at the junction. If the two pieces do not meet at the junction, then the function will have a discontinuity at the junction.

Example 2

Is the function $f(x) = \begin{cases} \frac{\sqrt{x+4}-2}{x} & x \neq 0 \\ \frac{1}{4} & x = 0 \end{cases}$ continuous at $x = 0$?

Solution: This is an example of a piecewise-defined function where the formula is used to evaluate the function for all values of x in the domain except one: $x = 0$. The point $x = 0$ is not in the domain of the piece $\frac{\sqrt{x+4}-2}{x}$: The denominator equals 0 when $x = 0$. Notice, however, that so does the numerator. In order for the function to be continuous at $x = 0$, $\lim_{x \to 0} \frac{\sqrt{x+4}-2}{x}$ must equal $\frac{1}{4}$. The technique to use to evaluate the limit is rationalizing the numerator:

Multiply the numerator and the denominator by the conjugate

$$\lim_{x \to 0} \frac{\sqrt{x+4}-2}{x} = \lim_{x \to 0} \frac{\sqrt{x+4}-2}{x} \cdot \frac{\left(\sqrt{x+4}+2\right)}{\left(\sqrt{x+4}+2\right)}$$

Simplify

$$\lim_{x \to 0} \frac{((x+4)-4)}{x\left(\sqrt{x+4}+2\right)}$$

Simplify and evaluate the limit using the substitution method

$$\lim_{x \to 0} \frac{\cancel{x}}{\cancel{x}\left(\sqrt{x+4}+2\right)} = \lim_{x \to 0} \frac{1}{\left(\sqrt{x+4}+2\right)} = \frac{1}{4}$$

Because $\lim_{x \to 0} \frac{\sqrt{x+4}-2}{x} = \frac{1}{4}$, we see that the limit exists.

The function is defined at $x = 0$: $f(0) = \frac{1}{4}$.

Thus the limit and the function are equal to each other at $x = 0$, and the function is continuous at $x = 0$.

Example 3

Is the function $f(x) = \begin{cases} x^2+1 & x<1 \\ 2x+1 & x \geq 1 \end{cases}$ continuous at $x = 1$?

Solution: This is an example of a piecewise-defined function whose domain is divided into two pieces. Each piece is a polynomial, which is as continuous as a function can get! The only possible discontinuity will occur at the junction, where the limit of the function may not exist. We will need to evaluate the limit as x approaches 1 from *below* using the first formula, and the limit as x approaches 1 from *above* using the second formula. If the limit from below is equal to the limit from above, this function has a shot at being continuous. If the two limits are not the same, then the function cannot be continuous at $x = 1$. We need to evaluate both limits individually. Because both pieces are continuous functions, we can use the substitution method to compute each limit:

$$\lim_{x \to 1^-} f(x) = \lim_{x \to 1^-} x^2 + 1 = (1)^2 + 1 = 2$$

$$\lim_{x \to 1^+} f(x) = \lim_{x \to 1^-} 2x + 1 = 2(1) + 1 = 3$$

Because the limit from below is not equal to the limit from above ($2 \neq 3$), the limit does not exist, and the function is not continuous at $x = 1$.

In the previous example, each piece of $f(x)$ was continuous (each piece was a polynomial), but the two pieces were not put together seamlessly. Just because each piece of a piecewise-defined function is continuous, that is not enough to make the overall function continuous. In order for a piecewise-defined function to be continuous, the pieces must be put together so that the limits from below and from above agree.

Lesson 6-1 Review

1. Evaluate $\lim_{x \to 2} \frac{x^3 + 4x^2 - 8x - 8}{x^3 - 2x^2 + x - 2}$.

2. Is the function $f(x) = \begin{cases} \frac{\sqrt{x+6}-3}{x-3} & x \neq 3 \\ \frac{1}{2} & x = 3 \end{cases}$ continuous at $x = 3$?

3. Is the function $f(x) = \begin{cases} x^2 + 1 & x < 1 \\ 3x - 1 & x \geq 1 \end{cases}$ continuous at $x = 1$?

Lesson 6-2 Types of Discontinuities

There are several ways in which a function can be discontinuous. Some discontinuities are more severe than others. As we saw with rational and trigonometric functions, a discontinuity can be associated with a vertical asymptote. In this case, the discontinuity cannot be patched or fixed. The function is heading towards infinity, and there is nothing we can do about it. Vertical asymptotes are examples of **non-removable discontinuities.**

Some types of discontinuities can be patched, or removed. The function $f(x)=\frac{x-1}{x^2-1}$ has a discontinuity at $x = 1$, but this discontinuity is not as severe as a vertical asymptote. The reason the discontinuity at $x = 1$ can be removed is because $\lim_{x\to 1}\frac{x-1}{x^2-1}$ exists. The discontinuity at $x = 1$ can be patched by defining the function at $x = 1$ separately. We can *define* $f(1)$ to be $\lim_{x\to 1}\frac{x-1}{x^2-1}$. This accomplishes two things: By defining the function at $x = 1$ we are putting 1 back into the domain of the function. Also, by defining $f(1)$ to be $\lim_{x\to 1}\frac{x-1}{x^2-1}$, we are guaranteeing that the third criterion for continuity is met. So, by carefully defining the function at $x = 1$ we are removing the discontinuity. Discontinuities that can be removed by redefining the function at that point are called **removable discontinuities.** They are also called removable singularities. One common place to find removable discontinuities is in a rational function where the numerator and denominator share common factors. We have seen other examples of removable discontinuities. The function $f(x)=\frac{\sin x}{x}$ has a removable discontinuity at $x = 0$. In Chapter 5 we evaluated $\lim_{x\to 0}\frac{\sin x}{x}$ numerically: $\lim_{x\to 0}\frac{\sin x}{x}=1$. The discontinuity at $x = 0$ cannot be eliminated through canceling common factors, but it can be eliminated by explicitly defining the function at $x = 0$:

$$f(x)=\begin{cases}\frac{\sin x}{x} & x\neq 0\\ 1 & x=0\end{cases}$$

Keep in mind that the material in calculus builds on previously developed skills and knowledge. We will put our knowledge of removing discontinuities to use when we study the derivative.

Discontinuities can also arise when we have a piecewise-defined function. If the two pieces of the function do not meet up at a single point, we have a **jump discontinuity**. As we saw in Chapter 5, the function

$$f(x)=\begin{cases} x^2+1 & x<1 \\ 2x+1 & x\geq 1 \end{cases}$$

has a jump discontinuity at $x=1$. We can see why this type of discontinuity is called a jump discontinuity by looking at the graph of this function, as shown in Figure 6.1. It looks as if the function value jumps from 2 to 3 at $x=1$. The limit from below and the limit from above are different, which is why the function is discontinuous. If the two limits were the same, the function would be continuous everywhere.

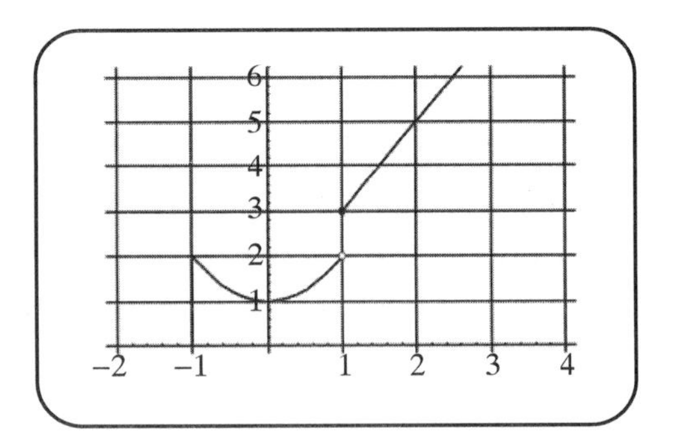

Figure 6.1.

Example 1

Find the real number k, which makes the function

$$f(x)=\begin{cases} 2x+1 & x<-1 \\ x^2+k & x\geq -1 \end{cases}$$ continuous everywhere.

Solution: Polynomials are continuous everywhere, so each piece of $f(x)$ is continuous, and the only place where $f(x)$ could possibly be discontinuous is where the two pieces fit together. If $\lim_{x\to-1^-} f(x) = \lim_{x\to-1^+} f(x)$, then the two pieces join up nicely and the function will be continuous. If we want this function to be continuous, we must choose a value for k so that the limit from below and the limit from above are equal:

$$\lim_{x\to-1^-} f(x) = \lim_{x\to-1^-} (2x+1) = 3$$

$$\lim_{x\to-1^+} f(x) = \lim_{x\to-1^+} \left(x^2+k\right) = 1+k$$

Set the two limits equal and solve for k:

$$3 = k + 1$$
$$k = 2$$

The function is continuous when $k = 2$.

Lesson 6-2 Review

1. Find c and k so that $f(x) = \begin{cases} x^2 & x < 3 \\ kx + c & 3 \le x \le 6 \\ x^2 & 6 < x \end{cases}$

   is continuous everywhere.

Lesson 6-3: The Intermediate Value Theorem

Part of calculus involves developing specific problem-solving strategies that can be applied in a variety of situations. For example, we have discussed several methods for evaluating the limit of a function. For each method there were specific conditions under which it should be used and a series of steps to take in order to solve a problem. These types of problem-solving strategies are sometimes referred to as "cookbook" mathematics, because there is a certain recipe that can be followed to successfully solve some types of problems.

Mathematics is much more than a collection of problem-solving recipes. Many results in mathematics have to do with determining whether or not a problem *can* be solved. No actual method is given for *how* to solve the problem. Establishing that certain problems *can* be solved is considered a mathematical result in and of itself. Mathematical results of this form are called **theorems**. The Intermediate Value Theorem is an example of this type of mathematical result.

The Intermediate Value Theorem

Suppose a function $f(x)$ is continuous on a closed interval $[a, b]$. If k is any number between $f(a)$ and $f(b)$, then there is at least one number c in the interval $[a, b]$ such that $f(c) = k$.

We can examine this theorem in more detail. Suppose that a continuous function starts out at the point $(a, f(a))$ and ends up at the point $(b, f(b))$, as shown in Figure 6.2. Then, if k is any number between $f(a)$ and $f(b)$, there is some point in the domain, call it c, that $f(x)$ maps on to k. In other words, the graph of $f(x)$ passes through the point (c, k). The key requirement for the Intermediate Value Theorem is continuity of the function $f(x)$. If you have to connect the points $(a, f(a))$ and $(b, f(b))$ in a continuous fashion (that is, without lifting a pencil when drawing the graph of the function), then $f(x)$ must pass through all of the points between $f(a)$ and $f(b)$ as x goes from a to b.

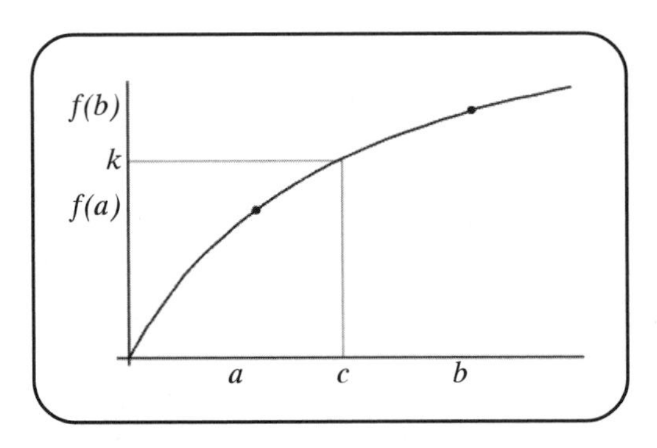

Figure 6.2.

The Intermediate Value Theorem may not appear to be very useful. It only states that *there is* a value c between a and b such that $f(c) = k$. It does not tell us how *many* points there are between a and b that map onto k, nor does it tell us *how* to find the location of one such point. The Intermediate Value Theorem is not a "cookbook" mathematical result. Surprisingly enough, though, it can be very useful.

One application of the Intermediate Value Theorem is in finding the zeros of a continuous function. Suppose that $f(x)$ is a continuous function, with $f(a) > 0$ and $f(b) < 0$. Then by the Intermediate Value Theorem, we know that the function takes on all values between $f(a)$ and $f(b)$. There is one number that is always between a negative number and a positive number: 0. By the Intermediate Value Theorem, there is a number, c, between a and b, satisfying $f(c) = 0$. In other words, the function has a root (or a zero) somewhere between a and b. The exact location of that root is not revealed by the Intermediate Value Theorem. The Intermediate Value Theorem only guarantees its existence.

There are several methods for finding a zero of a function in this situation. I will give one approach that is commonly used. Find the midpoint of the interval $[a, b]$, $m = \frac{1}{2}(a+b)$, and evaluate the function at $x = m$. If $f(m) = 0$, then you have found a zero! If $f(m) > 0$, then use the Intermediate Value Theorem with the interval $[m, b]$. If $f(m) < 0$, then use the Intermediate Value Theorem with the interval $[a, m]$. Keep dividing the interval in half and applying the Intermediate Value Theorem. This method of finding zeros is called the **Interval Bisection Method**. The *reason* it works is the Intermediate Value Theorem.

Example 1

Use the Intermediate Value Theorem to show that

$f(x) = x^4 + 2x - 1$ has a zero between $x = 0$ and $x = 1$.

Solution: Notice that the function is a polynomial, and polynomials are continuous on their domain, which is the set of all real numbers. So the criteria for the Intermediate Value Theorem is satisfied.

Also, $f(0) = -1$ and $f(1) = 2$, and 0 is between -1 and 2. Because the function is continuous and changes sign on $[0, 1]$, the Intermediate Value Theorem says that there is a number c between 0 and 1 satisfying $f(c) = 0$.

Example 2

The function $f(x)=\frac{2}{x-2}$ passes through the points (1, –2) and (3, 2). Because $f(1) < 0$ and $f(3) > 0$, by the Intermediate Value Theorem we would expect the function to have a zero between $x = 1$ and $x = 3$. In fact, $f(x)=\frac{2}{x-2}$ has no zeros. Is the Intermediate Value Theorem wrong?

Solution: The Intermediate Value Theorem is not wrong; it is being used incorrectly, as it does not apply to this situation. The function $f(x)=\frac{2}{x-2}$ is not continuous on [1, 3]: It has a vertical asymptote at $x = 2$.

Because the function is not continuous on [1, 3], the Intermediate Value Theorem does not apply.

The thing about theorems is that you have to make sure that you apply them correctly. The theorems apply only when the criteria are met. Be careful to read theorems carefully. It's important to understand what the theorems say *and* what they do not say!

Answer Key

Lesson 6-1 Review

1. $\lim\limits_{x\to 2}\frac{x^3+4x^2-8x-8}{x^3-2x^2+x-2}=\lim\limits_{x\to 2}\frac{\cancel{(x-2)}(x^2+6x+4)}{\cancel{(x-2)}(x^2+1)}=\lim\limits_{x\to 2}\frac{(x^2+6x+4)}{(x^2+1)}=\frac{20}{5}=4$

2. $\lim\limits_{x\to 3}\frac{\sqrt{x+6}-3}{x-3}=\lim\limits_{x\to 3}\frac{\sqrt{x+6}-3}{x-3}\cdot\frac{\sqrt{x+6}+3}{\sqrt{x+6}+3}=\lim\limits_{x\to 3}\frac{(x+6)-9}{(x-3)(\sqrt{x+6}+3)}=\lim\limits_{x\to 3}\frac{\cancel{(x-3)}}{\cancel{(x-3)}(\sqrt{x+6}+3)}=\frac{1}{6}$

   $\lim\limits_{x\to 3}\frac{\sqrt{x+6}-3}{x-3}=\frac{1}{6}$ and $f(3)=\frac{1}{2}$, which means that $\lim\limits_{x\to 3}\frac{\sqrt{x+6}-3}{x-3}\neq f(3)$, and the function is not continuous at $x=3$.

3. $\lim\limits_{x\to 1^-}f(x)=\lim\limits_{x\to 1^-}x^2+1=2$ and $\lim\limits_{x\to 1^+}f(x)=\lim\limits_{x\to 1^-}3x-1=2$, so $\lim\limits_{x\to 1}f(x)=2$

   $f(1)=2$, and $\lim\limits_{x\to 1}f(x)=f(1)$, so the function is continuous at $x=1$.

Lesson 6-2 Review

1. $\lim\limits_{x\to 3^-} f(x) = \lim\limits_{x\to 3^-} x^2 = 9$ and $\lim\limits_{x\to 3^+} f(x) = \lim\limits_{x\to 3^-} kx + c = 3k + c$.

   For the function to be continuous everywhere, $3k + c = 9$.

   $\lim\limits_{x\to 6^-} f(x) = \lim\limits_{x\to 6^-} kx + c = 6k + c$ and $\lim\limits_{x\to 6^+} f(x) = \lim\limits_{x\to 6^+} x^2 = 36$.

   For the function to be continuous everywhere, $6k + c = 36$.

   We have a system of equations that we can solve: $\begin{cases} 3k + c = 9 \\ 6k + c = 36 \end{cases}$.

   The solution to this system of equations is $k = 9$ and $c = -18$.

The Derivative

As I said before, calculus can be thought of as the study of limits. In the last chapter, we defined continuity in terms of limits. We will see that the definition of the derivative also involves limits.

The derivative is a sophisticated way to analyze a function. In algebra, the first method for graphing a function involves plotting points. After a while, plotting points becomes tiring, and new methods for graphing a function are introduced. For example, instead of plotting points, quadratic functions can be graphed using five characteristics: concavity, the axis of symmetry, the vertex, the y-intercept, and the x-intercept. The amount of work required to graph a quadratic function is decreased significantly. Using calculus, we will be able to analyze complicated functions more efficiently.

Lesson 7-1: Secant Lines and Difference Quotients

If a function $f(x)$ passes through the points $(a, f(a))$ and $(b, f(b))$, the **average rate of change** of the function between $x = a$ and $x = b$ is given by the formula:

$$\text{average rate of change} = \frac{f(b)-f(a)}{b-a}$$

The average rate of change of a function gives a general idea of how a function is changing over a specific interval. The function could undergo several changes in direction between $x = a$ and $x = b$. The average rate of change will not reveal any of those changes in direction.

This formula should look familiar: The average rate of change of a function over $[a, b]$ is just the *slope* of the line that passes through the points $(a, f(a))$ and $(b, f(b))$:

$$\text{slope} = \frac{f(b)-f(a)}{b-a}$$

A line that passes through two points of the graph of a function is called a **secant** line, and the slope of the secant line between $(a, f(a))$ and $(b, f(b))$ is the average rate of change of the function $f(x)$ between $x = a$ and $x = b$.

Example 1

Find the average rate of change of the function $f(x) = x^2$ between $x = 2$ and $x = 4$, and find the equation of the secant line passing through $(2, f(2))$ and $(4, f(4))$.

Solution: Using the equation for the average rate of change, we have:

$$\text{average rate of change} = \frac{f(b)-f(a)}{b-a}$$

Substitute $a = 2$ and $b = 4$ into the equation for the average rate of change

$$\text{average rate of change} = \frac{f(4)-f(2)}{4-2}$$

$f(2) = 2^2 = 4$, and $f(4) = 4^2 = 16$

$$\text{average rate of change} = \frac{16-4}{2}$$

Simplify

$$\text{average rate of change} = 6$$

To find the equation of the secant line, substitute the slope (the average rate of change that we just calculated) and one of the points into the point-slope formula for a line: $y - 4 = 6(x - 2)$

$$y = 6x - 8.$$

There are several formulas that can be used to find the slope of the secant line, but they all say basically the same thing. To find the slope of the secant line of the function $f(x)$ over the interval $[x, x + h]$ we evaluate:

$$\text{slope} = \frac{f(x+h)-f(x)}{(x+h)-x} = \frac{f(x+h)-f(x)}{h}$$

We discussed this formula in Chapter 1. We even gave it a name: It was called the difference quotient. I assured you in Lesson 1-6 that this formula would be important in calculus. It is the basis for the derivative. We will be working with the difference quotient throughout this chapter.

Example 2

Find the average rate of change of the function $f(x) = 3x^2$ between $x = 1$ and $x = 1 + h$, where $h \neq 0$, in terms of h.

Solution: Using the equation for the average rate of change, we have:

$$\text{average rate of change} = \frac{f(b)-f(a)}{b-a}$$

Substitute $a = 1$ and $b = 1 + h$ into the equation for the average rate of change:

$$\text{average rate of change} = \frac{f(1+h)-f(1)}{(1+h)-1}$$

$f(1 + h) = 3(1 + h)^2$, and $f(1) = 3 \cdot 1^2 = 3$:

$$\text{average rate of change} = \frac{3(1+h)^2 - 3}{h}$$

Expand $3(1 + h)^2$:

$$\text{average rate of change} = \frac{3(1+2h+h^2)-3}{h}$$

Combine the constant terms:

$$\text{average rate of change} = \frac{3+6h+3h^2-3}{h}$$

Factor an h out of each term in the numerator:

$$\text{average rate of change} = \frac{6h+3h^2}{h}$$

Cancel the h in the numerator with the h in the denominator, because $h \neq 0$:

$$\text{average rate of change} = \frac{\cancel{h}(6+3h)}{\cancel{h}} = (6+3h)$$

The difference quotient is the slope of the secant line passing through the points $(x, f(x))$ and $(x + h, f(x + h))$. The difference quotient is also the average rate of change of the function $f(x)$ over the interval $[x, x + h]$. The average rate of change gives an idea of the overall change of a function over an interval. Knowing the overall change of a function over an interval can certainly be useful, but it may be more useful to know how the function is changing at an *instant*.

Suppose $f(x)$ represents the position of an object as a function of time, x. Then $f(b) - f(a)$ represents the change in location, or the distance traveled, between time $x = a$ and time $x = b$, and the quantity $(b - a)$ represents the length of time that has elapsed. The average rate of change of $f(x)$ over the interval $[a, b]$, $\frac{f(b)-f(a)}{b-a}$, can then be interpreted as the ratio of the distance traveled divided by the time elapsed, or the average *velocity*.

Figure 7.1 will help us visualize secant lines more clearly. The function shown in Figure 7.1 passes through the points (0, 1) and (2, 5). The secant line is the line that passes through those two points. The average rate of change of the function over the interval [0, 2] is: $\frac{f(2)-f(0)}{2-0} = \frac{5-1}{2-0} = 2$. This is equivalent to saying that the slope of the secant line that passes through the points (0, 1) and (2, 5) is 2.

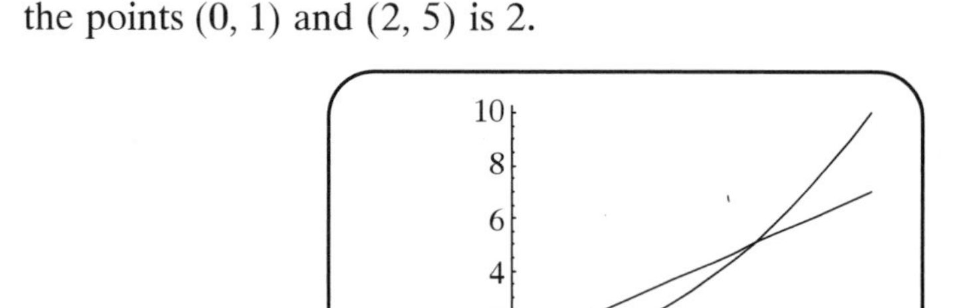

Figure 7.1.

Lesson 7-1 Review

1. Find the slope of the secant line of the given functions over the indicated intervals:
   a. $f(x) = 2^x$ over $[1, 3]$
   b. $g(x) = \sin x$ over $\left[0, \frac{\pi}{2}\right]$
   c. $h(x) = \log_3 x$ over $\left[\frac{1}{3}, 3\right]$

2. Find the average rate of change of the following functions over the given intervals:
   a. $g(x) = x^2 - 2$ over $[0, 2]$
   b. $f(x) = \tan x$ over $\left[0, \frac{\pi}{4}\right]$
   c. $h(x) = \log_2 x$ over $[1, 4]$

Lesson 7-2: Tangent Lines

Secant lines intersect a function at two distinct points, and the slope of the secant line represents the *average rate of change of a function* over a given interval. There are times when it is more useful to know the *instantaneous* rate of change of a function at a *point* $x = a$. Before we discuss what is meant by the instantaneous rate of change, we need to discuss tangent lines.

A line is **tangent** to a function at a point $x = a$ if the line just touches the graph of the function at that point. Figure 7.2 shows the function $f(x) = x^2$ with two of its tangent lines: one at $x = -1$ and the other at $x = 2$. Notice that the tangent lines just glance off of the graph of the function. In order to find the equation of a tangent line, you need to know the point where the tangent line touches the graph of the function and the slope of the tangent line. That's where calculus comes into play.

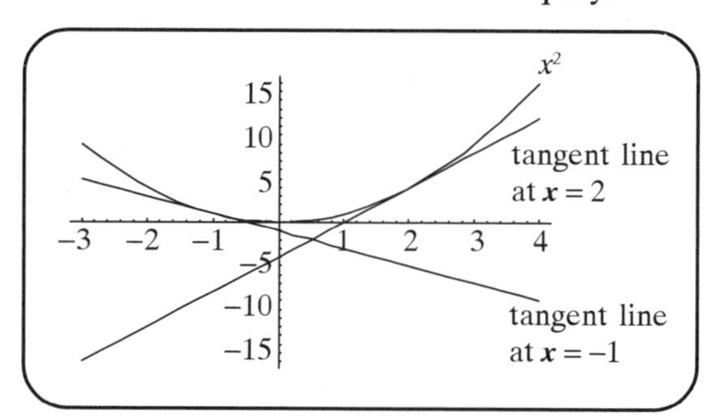

Figure 7.2.

The slope of the secant line passing through the points $(a, f(a))$ and $(a + h, f(a + h))$ is the ratio:

$$\frac{f(a+h)-f(a)}{(a+h)-a} = \frac{f(a+h)-f(a)}{h}$$

If h is small, then the secant line is very close to the tangent line. The smaller the value of h, the closer the secant line gets to the tangent line. If we let h $\to 0$, then the secant line $\to$ the tangent line, and the slope of the secant line approaches the slope of the tangent line:

$$\lim_{h\to 0}(\text{slope of secant line}) = \text{slope of tangent line}$$

If we denote the slope of the line tangent to the graph of the function $f(x)$ at $x = a$ by $f'(a)$, we have:

$$f'(a) = \lim_{h\to 0}\frac{f(a+h)-f(a)}{h}$$

We call $f'(a)$ the derivative of the function $f(x)$ at $x = a$. The **derivative** of a function at a point $x = a$ is the slope of the line tangent to the graph of $f(x)$ at $x = a$. The definition of the derivative of a function involves a limit. Remember that calculus is the study of limits. The notion of continuity was the first application of a limit. The derivative of a function is our second application of limits.

Lesson 7-3: The Definition of the Derivative

The derivative of a function at $x = a$ is the slope of the line tangent to the graph of $f(x)$ at $x = a$. The lines tangent to the graph of a function do not all have the same slope. Sometimes the tangent lines will have a positive slope, and other times the tangent lines will have a negative slope. The sign and the magnitude of the slope of the tangent lines depend on how the function is changing. The derivative of a function measures the rate of change of a function along its graph. In fact, the derivative of a function is itself a function. We define the function $f'(x)$ as:

$$f'(x) = \lim_{h\to 0}\frac{f(x+h)-f(x)}{h}$$

We can find the derivative of various functions by evaluating this limit using the techniques discussed in Chapter 5.

Example 1

Find the derivative of the function $f(x) = x^2$.

Solution: Use the definition of the derivative and evaluate the limit directly. If $f(x) = x^2$, then $f(x + h) = (x + h)^2 = x^2 + 2xh + h^2$. Substituting into the definition for the derivative, we have:

Start with the definition of the derivative $f'(x)=\lim_{h\to 0}\frac{f(x+h)-f(x)}{h}$

Substitute in for $f(x)$, and $f(x+h)$ $f'(x)=\lim_{h\to 0}\frac{(x^2+2xh+h^2)-x^2}{h}$

The x^2 terms subtract out $f'(x)=\lim_{h\to 0}\frac{2xh+h^2}{h}$

Cancel the h's $f'(x)=\lim_{h\to 0}\frac{\cancel{h}(2x+h)}{\cancel{h}}$

Take the limit. The term $2x$ is not affected by changes in h.

$$f'(x)=\lim_{h\to 0}(2x+h)$$

$$f'(x)=2x$$

The derivative of the function $f(x)=x^2$ is the function $f'(x)=2x$.

Example 2

Find the derivative of the function $f(x)=\sqrt{x}$.

Solution: Use the definition of the derivative and evaluate the limit directly. If $f(x)=\sqrt{x}$, then $f(x+h)=\sqrt{x+h}$. Substituting into the definition for the derivative, we have:

Start with the definition of the derivative

$$f'(x)=\lim_{h\to 0}\frac{f(x+h)-f(x)}{h}$$

Substitute in for $f(x)$ and $f(x+h)$ $f'(x)=\lim_{h\to 0}\frac{\sqrt{x+h}-\sqrt{x}}{h}$

Multiply by the conjugate of the numerator

$$f'(x)=\lim_{h\to 0}\frac{\sqrt{x+h}-\sqrt{x}}{h}\cdot\frac{\sqrt{x+h}+\sqrt{x}}{\sqrt{x+h}+\sqrt{x}}$$

The x's subtract out $f'(x)=\lim_{h\to 0}\frac{(x+h)-x}{h\left(\sqrt{x+h}+\sqrt{x}\right)}$

Cancel the h's $$f'(x)=\lim_{h\to 0}\frac{\cancel{h}}{\cancel{h}\left(\sqrt{x+h}+\sqrt{x}\right)}$$

Evaluate the limit; x is not affected by changes in h

$$f'(x)=\lim_{h\to 0}\frac{1}{\left(\sqrt{x+h}+\sqrt{x}\right)}$$

$$f'(x)=\frac{1}{2\sqrt{x}}$$

The derivative of the function $f(x)=\sqrt{x}$ is the function $f'(x)=\frac{1}{2\sqrt{x}}$.

Example 3

Find the derivative of the function $f(x) = e^x$.

Solution: To solve this problem, we need to make use of a previous result: $\lim_{x\to 0}\frac{e^x-1}{x}=1$. Keep in mind that the variable used in the limit expression is not important; any letter of the alphabet would serve the same purpose as long as one variable is consistently replaced with another variable throughout the limit expression. This equation can also be written $\lim_{h\to 0}\frac{e^h-1}{h}=1$. We will make use of this limit in evaluating the limit of the difference quotient for the function $f(x) = e^x$. First, evaluate $f(x + h)$: $f(x + h) = e^{x+h}$. Substitute into the definition of the derivative and evaluate the limit.

Start with the definition of the derivative

$$f'(x)=\lim_{h\to 0}\frac{f(x+h)-f(x)}{h}$$

Substitute in for $f(x)$ and $f(x + h)$ $$f'(x)=\lim_{h\to 0}\frac{e^{x+h}-e^x}{h}$$

Factor e^x out of each expression in the numerator

$$f'(x)=\lim_{h\to 0}\frac{e^x\left(e^h-1\right)}{h}$$

The term e^x has no dependence on h, and it is not affected by the behavior of h. It can be factored out of the limit expression.

$$f'(x) = e^x \lim_{h \to 0} \frac{(e^h - 1)}{h}$$

Evaluate the limit $\qquad f'(x) = e^x \cdot 1$

The derivative of the function $f(x) = e^x$ is $f'(x) = e^x$.

Example 4

Find the derivative of the function $f(x) = \sin x$.

Solution: Start with the definition of the derivative and simplify. In this situation, the sum of two angles formula will come in handy. We will also make use of two results obtained in Chapter 5: $\lim_{x \to 0} \frac{\sin x}{x} = 1$ and $\lim_{x \to 0} \frac{\cos x - 1}{x} = 0$. These two results are instrumental in determining the derivative of the sine function.

Start with the definition of the derivative

$$f'(x) = \lim_{h \to 0} \frac{f(x+h) - f(x)}{h}$$

Substitute in for $f(x)$ and $f(x + h)$

$$f'(x) = \lim_{h \to 0} \frac{\sin(x+h) - \sin x}{h}$$

Use the sum formula to expand $\sin(x + h)$

$$f'(x) = \lim_{h \to 0} \frac{\sin x \cos h + \sin h \cos x - \sin x}{h}$$

Regroup the numerator $\qquad f'(x) = \lim_{h \to 0} \frac{(\sin x \cos h - \sin x) + \sin h \cos x}{h}$

Split the numerator into two pieces

$$f'(x) = \lim_{h \to 0} \left(\frac{\sin x \cos h - \sin x}{h} + \frac{\sin h \cos x}{h} \right)$$

The $\sin x$ and $\cos x$ are not affected by h, so they can be factored out of the limit $\qquad f'(x) = \lim_{h \to 0} \left((\sin x) \frac{(\cos h - 1)}{h} + (\cos x) \frac{\sin h}{h} \right)$

Evaluate each limit separately $f'(x) = (\sin x)\lim\limits_{h\to 0}\frac{(\cos h - 1)}{h} + (\cos x)\lim\limits_{h\to 0}\frac{\sin h}{h}$

$$f'(x) = (\sin x)\cdot 0 + (\cos x)\cdot 1$$

$$f'(x) = \cos x$$

The derivative of the function $f(x) = \sin x$ is $f'(x) = \cos x$.

Lesson 7-3 Review

Use the definition of the derivative (the difference quotient) to find the derivative of the following functions:

1. $f(x) = x^2 + 3x$
2. $f(x) = \sqrt{x+2}$
3. $f(x) = e^{x+3}$
4. $f(x) = \frac{1}{x}$
5. $g(x) = \cos x$

Lesson 7-4: The Existence of the Derivative

In order for the derivative of a function $f(x)$ at $x = a$ to exist, $\lim\limits_{h\to 0}\frac{f(a+h)-f(a)}{h}$ must exist. The denominator of the difference quotient is heading towards 0. The only way for this limit to exist is if the *numerator* also heads towards 0. The only way for the numerator to head towards 0 is if $\lim\limits_{h\to 0} f(a+h) = f(a)$. This means that the function $f(x)$ is *continuous* at $x = a$. In other words, if the derivative of a function at a point $x = a$ exists, then at a *minimum*, the function must be continuous at $x = a$. In other words, the derivative of a function cannot be defined wherever the function has holes, jump discontinuities, or vertical asymptotes.

The requirement that a function be continuous is, as I mentioned, a minimum requirement. There are functions that are continuous on their domain that have points where the derivative does not exist. The function $f(x) = |x|$ is an example of a function that is continuous everywhere, yet its derivative does not exist at the point $x = 0$. The graph of the function $f(x) = |x|$ is shown in Figure 7.3 on page 123.

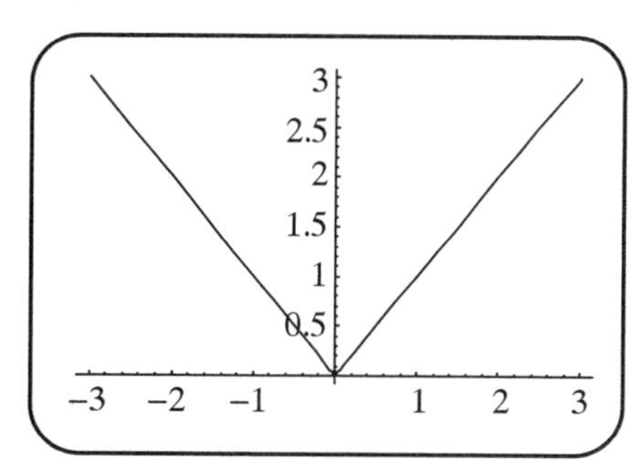

Figure 7.3.

To understand why the derivative does not exist at $x = 0$, we will try to evaluate $f'(0)$ using the difference quotient. Recall that the absolute value function is a piecewise-defined function and can be written:

$$f(x) = |x| = \begin{cases} -x & x < 0 \\ x & x \geq 0 \end{cases}$$

To evaluate the derivative of this function at $x = 0$, we will need to evaluate:

$$\lim_{h \to 0} \frac{f(0+h) - f(0)}{h} = \lim_{h \to 0} \frac{|h| - |0|}{h} = \lim_{h \to 0} \frac{|h|}{h}$$

In order to evaluate this limit, we will need to let $h \to 0$ from below and from above:

$$\lim_{h \to 0^+} \frac{|h|}{h} = \lim_{h \to 0^+} \frac{h}{h} = 1$$

$$\lim_{h \to 0^-} \frac{|h|}{h} = \lim_{h \to 0^+} \frac{-h}{h} = -1$$

Because the limit from below and the limit from above are different, we end up with two values for the derivative at $x = 0$: 1 and –1. It does not make sense to have the slope of the tangent line flipping between 1 and –1. Our only other option is to say that the derivative of the function $f(x) = |x|$ does not exist at $x = 0$.

The absolute value function is continuous everywhere, but its derivative does not exist at $x = 0$. Notice that the graph of $f(x) = |x|$ has a sharp corner at $x = 0$. Sharp corners and cusps are graphical indicators of problems with the derivative. Algebraically, functions that are defined piecewise may have problems with the derivative where the pieces meet. In order to determine whether or not the derivative of a piecewise-defined function exists at the seam, you must use the definition of the derivative and evaluate the limit from above and from below. If the two results are the same, then the derivative exists. Otherwise, the derivative does not.

Another problem with the existence of the derivative occurs if the tangent line is vertical. The derivative is the slope of the tangent line, and the slope of a vertical line is undefined. So if the graph of a function has a vertical tangent line at $x = a$, then $f'(a)$ will not be defined. An example of a function with a vertical tangent line is $f(x)=\sqrt[3]{x}$. The graph of $f(x)=\sqrt[3]{x}$ is shown in Figure 7.4. Notice that at $x = 0$, the tangent line is vertical: $f'(0)$ is not defined.

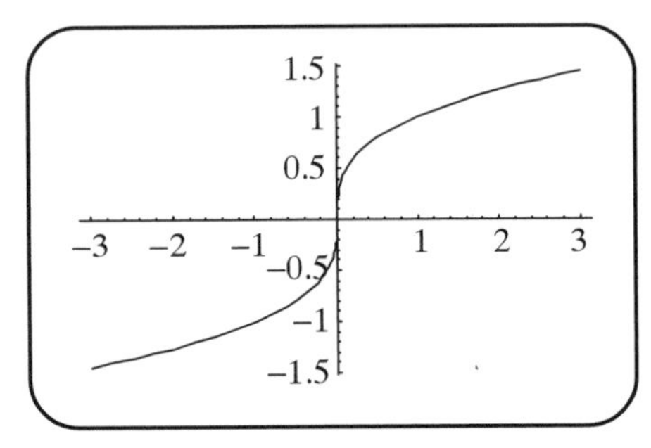

Figure 7.4.

Example 1

Does the derivative of the function $g(x)=\begin{cases} x & x<1 \\ \sqrt{x} & x\geq 1 \end{cases}$ exist at $x = 1$?

Solution: We need to evaluate $\lim_{h\to 0^+}\frac{g(1+h)-g(1)}{h}$ and $\lim_{h\to 0^-}\frac{g(1+h)-g(1)}{h}$ and compare the results.

Start with the definition of the derivative $\lim_{h\to 0^+}\frac{g(1+h)-g(1)}{h}$

If $h > 0$, then $1 + h > 1$ and we use the second formula for $g(x)$

$$\lim_{h\to 0^+}\frac{\sqrt{1+h}-\sqrt{1}}{h}$$

Multiply both the numerator and denominator by the conjugate

$$\lim_{h\to 0^+}\frac{\sqrt{1+h}-\sqrt{1}}{h}\cdot\frac{\sqrt{1+h}+\sqrt{1}}{\sqrt{1+h}+\sqrt{1}}$$

Simplify $\lim_{h\to 0^+} \frac{(1+h)-1}{h\left(\sqrt{1+h}+\sqrt{1}\right)}$

Cancel the h's $\lim_{h\to 0^+} \frac{\cancel{h}}{\cancel{h}\left(\sqrt{1+h}+\sqrt{1}\right)}$

Evaluate the limit $\lim_{h\to 0^+} \frac{1}{\left(\sqrt{1+h}+\sqrt{1}\right)}$

$$\lim_{h\to 0^+} \frac{1}{\left(\sqrt{1+h}+\sqrt{1}\right)} = \frac{1}{2}$$

Next, evaluate the limit from below:

Start with the definition of the derivative $\lim_{h\to 0^-} \frac{g(1+h)-g(1)}{h}$

If $h < 0$, then $1 + h < 1$ and we use the first formula for $g(x)$. We still use $g(1)=\sqrt{1}$, however. $\lim_{h\to 0^-} \frac{(1+h)-\sqrt{1}}{h}$

Simplify $\lim_{h\to 0^-} \frac{(1+h)-1}{h}$

Simplify $\lim_{h\to 0^-} \frac{h}{h} = 1$

The limit from below and the limit from above are not equal to each other. The derivative from above is $\frac{1}{2}$ and the derivative from below is 1, which means that the derivative does not exist. The graph of $g(x)$ is shown in Figure 7.5. Even though the pieces of the function $g(x)$ line up to give a continuous function, the derivative does not exist at $x = 1$.

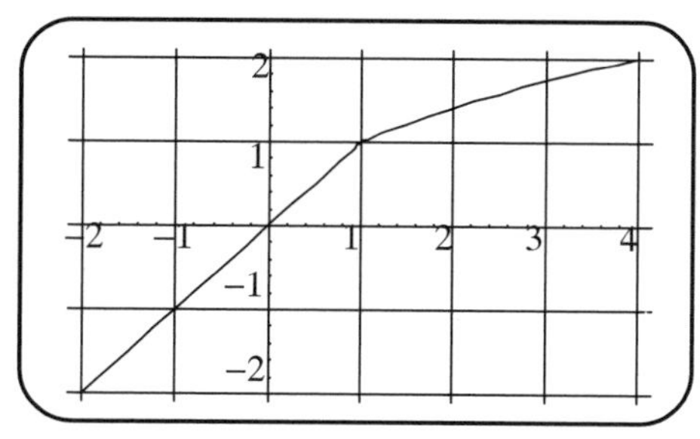

Figure 7.5.

Lesson 7-4 Review

Use the definition of the derivative (the difference quotient) to find the derivative of the following functions:

1. Does the derivative of the function $f(x)=\begin{cases} x^2+1 & x<0 \\ 2x+1 & x\geq 0 \end{cases}$ exist at $x = 0$?

2. Does the derivative of the function $f(x)=\begin{cases} x^2 & x<0 \\ x^3 & x\geq 0 \end{cases}$ exist at $x = 0$?

Lesson 7-5: The Derivative and Tangent Line Equations

To find the equation of a line, we need two things: a point and a slope. When finding the equation of the line tangent to the graph of $f(x)$ at $x = a$, the point that the line passes through is $(a, f(a))$. The slope of the line tangent to the graph of $f(x)$ at $x = a$ is $f'(a)$. Using these two pieces of information, we can use the point-slope equation for a line:

$$y - f(a) = f'(a)(x - a)$$

Realize that the quantities a, $f(a)$ and $f'(a)$ are all *numbers*. The equation of a tangent line must actually be the equation of a line!

Example 1

Find the equation of the line tangent to the graph of $f(x) = x^2 + 1$ at $x = 3$. Write your answer in slope-intercept form.

Solution: We need a point and a slope. The point is $(3, f(3))$, or $(3, 10)$. The slope is $f'(3)$, which we will find using the definition of the derivative:

Start with the definition of the derivative $f'(3)=\lim\limits_{h\to 0}\dfrac{f(3+h)-f(3)}{h}$

Substitute in for $f(3 + h)$ and $f(3)$ $f'(3)=\lim\limits_{h\to 0}\dfrac{\left((3+h)^2+1\right)-10}{h}$

Expand $(3 + h)^2$ $f'(3)=\lim\limits_{h\to 0}\dfrac{\left(9+6h+h^2+1\right)-10}{h}$

Simplify $f'(3)=\lim\limits_{h\to 0}\dfrac{\left(h^2+6h+10\right)-10}{h}$

Factor an h out of both terms in the numerator

$$f'(3)=\lim_{h\to 0}\frac{h^2+6h}{h}$$

Cancel the h's

$$f'(3)=\lim_{h\to 0}\frac{\not{h}(h+6)}{\not{h}}$$

Evaluate the limit

$$f'(3)=\lim_{h\to 0}(h+6)=6$$

The slope of the tangent line is 6. Now we can substitute into the point-slope formula for a line and simplify:

The point is (3, 10) and the slope is 6 $\quad y - 10 = 6(x - 3)$

Simplify

$$y-10=6x-18$$
$$y=6x-8$$

The equation of the line tangent to the graph of $f(x) = x^2 + 1$ at $x = 3$ is $y = 6x - 8$.

Figure 7.6 shows the graphs of $f(x) = x^2 + 1$ and the line $y = 6x - 8$. Notice that the line just touches the graph of $f(x) = x^2 + 1$ at $x = 3$.

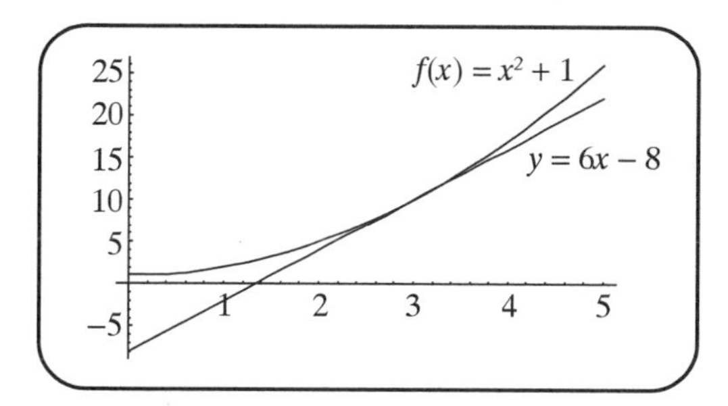

Figure 7.6.

Example 2

Find the equation of the line tangent to the graph of $f(x) = e^x$ at $x = 0$. Write your answer in slope-intercept form.

Solution: In order to find the equation of a tangent line, we need a point and a slope. The point is $(0, f(0))$, or $(0, 1)$. We can use Example 3 of Lesson 7-3 to find the slope of the tangent line. We derived a formula for the derivative of $f(x) = e^x$: $f'(x) = e^x$. We can use this formula to find the derivative (or the slope of the tangent line) for any value of x: $f'(0) = e^0 = 1$.

So the slope is 1, and the equation of the line tangent to the graph of $f(x) = e^x$ at $x = 0$ is:

$$y - 1 = 1(x - 0)$$
$$y = x + 1$$

Figure 7.7 shows the graph of the function $f(x) = e^x$ and the line $y = x + 1$. Notice that the line just touches the graph of $f(x) = e^x$ at $x = 0$.

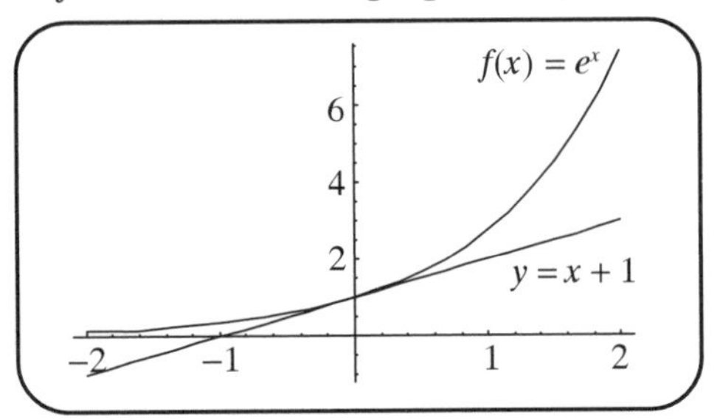

Figure 7.7.

Example 3

Find the equation of the line tangent to the graph of $g(x) = \sin x$ at $x = 0$. Write your answer in slope-intercept form.

Solution: In order to find the equation of a tangent line, we need a point and a slope. The point is $(0, g(0))$, or $(0, 0)$. We can use Example 4 of Lesson 7-3 to find the slope of the tangent line. We derived a formula for the derivative of $g(x) = \sin x$: $g'(x) = \cos x$. We can use this formula to find the derivative (or the slope of the tangent line) for any value of x: $g'(0) = \cos 0 = 1$. So the slope is 1, and the equation of the line tangent to the graph of $g(x) = \sin x$ at $x = 0$ is:

$$y - 0 = 1(x - 0)$$
$$y = x$$

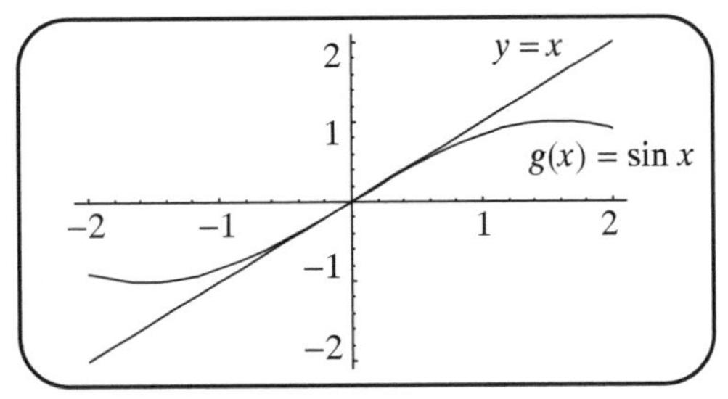

Figure 7.8.

Figure 7.8 shows the graph of the function $g(x) = \sin x$ and the line $y = x$. Notice that the line just touches the graph of $g(x) = \sin x$ at $x = 0$.

Example 4

Find the equation of the line tangent to the graph of $f(x)=\sqrt{x}$ at $x = 4$. Write your answer in slope-intercept form.

Solution: In order to find the equation of a tangent line, we need a point and a slope. The point is $(4, f(4))$, or $(4, 2)$. We can use Example 2 of Lesson 7-3 to find the slope of the tangent line. We derived a formula for the derivative of $f(x)=\sqrt{x}$: $f'(x)=\frac{1}{2\sqrt{x}}$. We can use this formula to find the derivative (or the slope of the tangent line) for any value of x: $f'(4)=\frac{1}{2\sqrt{4}}=\frac{1}{4}$. So the slope is $\frac{1}{4}$, and the equation of the line tangent to the graph of $f(x)=\sqrt{x}$ at $x = 4$ is:

$$y-2=\tfrac{1}{4}(x-4)$$
$$y-2=\tfrac{1}{4}x-1$$
$$y=\tfrac{1}{4}x+1$$

Figure 7.9 shows the graph of the function $f(x)=\sqrt{x}$ and the line $y=\frac{1}{4}x+1$. Notice that the line just touches the graph of $f(x)=\sqrt{x}$ at $x=4$.

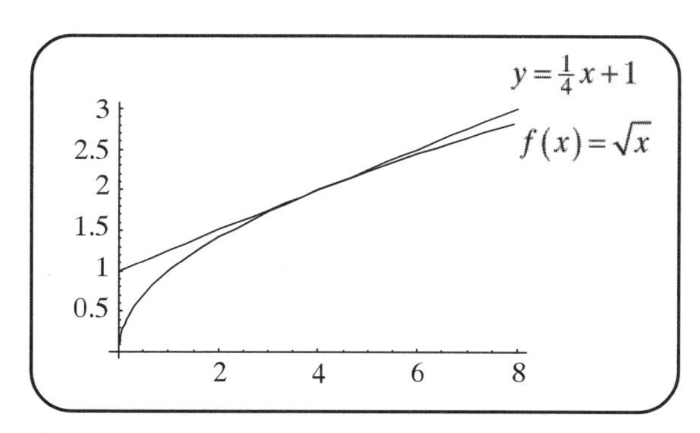

Figure 7.9.

Lesson 7-5 Review

Use the Lesson 7-3 Review problems to find the equation of the line tangent to the graphs of the following functions at the point indicated:

1. $f(x)=x^2+3x,\ x=2$
2. $f(x)=\sqrt{x+2},\ x=2$
3. $f(x)=e^{x+3},\ x=-3$
4. $f(x)=\frac{1}{x},\ x=1$
5. $g(x)=\cos x,\ x=\frac{\pi}{2}$

Lesson 7-6: Notation

Calculus has developed over the years, and, as calculus has changed, so has its notation. I have been using $f'(x)$ to denote the derivative of the function $f(x)$. We usually write $f'(a)$ when we talk about the derivative of the function $f(x)$ at the point $x = a$. We can also write the derivative of the function $f(x)$ with respect to x as $\frac{df}{dx}$. This notation is called **Leibniz notation**, named after one of the mathematicians credited with developing calculus. The phrase "with respect to x" means that the independent variable in our function is x. This notation suggests that the derivative is a fraction. Because the derivative is a limit of a difference quotient, this notation follows naturally. If we write $\Delta f = f(x+h) - f(x)$, and $\Delta x = (x+h) - x$, then we can write the derivative as:

$$\frac{df}{dx} = \lim_{\Delta x \to 0} \frac{\Delta f}{\Delta x}$$

The derivative of the function $f(x) = x$ with respect to x can be thought of as $\frac{dx}{dx}$, and if we pretend that this is a fraction, we see that the numerator and the denominator are the same thing, so we would expect this value to be 1. In fact, we can use the definition of the derivative to show that the derivative of the function $f(x) = x$ with respect to x is, in fact, 1:

$$\lim_{h \to 0} \frac{f(x+h)-f(x)}{h} = \lim_{h \to 0} \frac{(x+h)-x}{h} = \lim_{h \to 0} \frac{h}{h} = 1$$

If we want to express the derivative of the function $f(x)$ at the point $x = a$ using Leibniz notation, we write $\left.\frac{df}{dx}\right|_{x=a}$. In other words, $f'(a) = \left.\frac{df}{dx}\right|_{x=a}$.

But wait! There's more! If we write $y = f(x)$, then the derivative of $f(x)$ with respect to x can also be written as y' or $\frac{dy}{dx}$. If our function is actually a function of time, so that t is the independent variable instead of x, then we would be taking the derivative with respect to t, and we would write $f'(t)$, $\frac{dx}{dt}$, y' or $\frac{dy}{dt}$.

To make matters worse, physicists also use dot notation to represent the derivative with respect to time: $\dot{y} = \frac{dy}{dt}$. Fortunately, they reserve this notation for derivatives with respect to time only, so there is no confusion between y' and $\dot{y}$. We also write $f'(x)$ as $\frac{d}{dx}\left[f(x)\right]$, $D_x[f(x)]$, and $D_x y$. The subscript x specifies the independent variable, or the variable that the derivative is taken with respect to. Keep in mind that these variables are "dummy" variables, meaning that these variables can be interchanged. If, for whatever reason, y is the independent variable, then we could talk about $f'(y)$, or $\frac{df}{dy}$. Leibniz notation helps keep track of the players: which is the independent variable, and which variable is dependent.

The notation for the derivative has evolved over the years, and everyone has a favorite way to represent the derivative. There are times when one particular type of notation works better than another, which is why there are so many different symbols for the derivative.

Answer Key

Lesson 7-1 Review

1. a. $f(x) = 2^x$ over $[1, 3]$: $\frac{f(3)-f(1)}{3-1} = \frac{8-2}{2} = 3$

   b. $g(x) = \sin x$ over $\left[0, \frac{\pi}{2}\right]$: $\frac{g\left(\frac{\pi}{2}\right)-g(0)}{\frac{\pi}{2}-0} = \frac{\sin\frac{\pi}{2}-\sin 0}{\frac{\pi}{2}} = \frac{1}{\frac{\pi}{2}} = \frac{2}{\pi}$

   c. $h(x) = \log_3 x$ over $\left[\frac{1}{3}, 3\right]$: $\frac{h(3)-h\left(\frac{1}{3}\right)}{3-\frac{1}{3}} = \frac{\log_3 3 - \ln\frac{1}{3}}{\frac{8}{3}} = \frac{1-(-1)}{\frac{8}{3}} = \frac{2}{\frac{8}{3}} = \frac{3}{4}$

2. a. $g(x) = x^2 - 2$ over $[0, 2]$: $\frac{g(2)-g(0)}{2-0} = \frac{2-(-2)}{2} = 2$

b. $f(x) = \tan x$ over $\left[0, \frac{\pi}{4}\right]$: $\frac{f\left(\frac{\pi}{4}\right)-f(0)}{\frac{\pi}{4}-0} = \frac{\tan\frac{\pi}{4}-\tan 0}{\frac{\pi}{4}} = \frac{1}{\frac{\pi}{4}} = \frac{4}{\pi}$

c. $h(x) = \log_2 x$ over $[1, 4]$: $\frac{h(4)-h(1)}{4-1} = \frac{\log_2 4-\log_2 1}{3} = \frac{2-0}{3} = \frac{2}{3}$

Lesson 7-3 Review

1. $\lim\limits_{h\to 0}\frac{f(x+h)-f(x)}{h} = \lim\limits_{h\to 0}\frac{(x+h)^2+3(x+h)-(x^2+3x)}{h} = 2x+3$

2. $\lim\limits_{h\to 0}\frac{f(x+h)-f(x)}{h} = \lim\limits_{h\to 0}\frac{\sqrt{x+h+2}-\sqrt{x+2}}{h} = \lim\limits_{h\to 0}\frac{\sqrt{x+h+2}-\sqrt{x+2}}{h}\cdot\frac{\sqrt{x+h+2}+\sqrt{x+2}}{\sqrt{x+h+2}+\sqrt{x+2}} = \frac{1}{2\sqrt{x+2}}$

3. $\lim\limits_{h\to 0}\frac{f(x+h)-f(x)}{h} = \lim\limits_{h\to 0}\frac{e^{x+h+3}-e^{x+3}}{h} = \lim\limits_{h\to 0}\frac{e^{x+3}\left(e^h-1\right)}{h} = e^{x+3}$

4. $\lim\limits_{h\to 0}\frac{f(x+h)-f(x)}{h} = \lim\limits_{h\to 0}\frac{\frac{1}{x+h}-\frac{1}{x}}{h} = \lim\limits_{h\to 0}\frac{\frac{x}{x(x+h)}-\frac{(x+h)}{x(x+h)}}{h} = \lim\limits_{h\to 0}\frac{-h}{hx(x+h)} = -\frac{1}{x^2}$

5. Use the sum of two angles formula for the cosine function:

$$\lim_{h\to 0}\frac{f(x+h)-f(x)}{h} = \lim_{h\to 0}\frac{\cos(x+h)-\cos x}{h} = \lim_{h\to 0}\frac{\cos x\cos h-\sin x\sin h-\cos x}{h}$$

$$= \lim_{h\to 0}\left[(\cos x)\frac{(\cos h-1)}{h}-(\sin x)\frac{\sin h}{h}\right]$$

$$= (\cos x)\cdot 0-(\sin x)\cdot 1 = -\sin x$$

Lesson 7-4 Review

1. Evaluate the derivative using the definition:

$$f'(0) = \lim_{h\to 0^-}\frac{f(0+h)-f(0)}{h} = \lim_{h\to 0^-}\frac{\left(h^2+1\right)-(1)}{h} = \lim_{h\to 0^-}\frac{h^2}{h} = 0$$

$$f'(0) = \lim_{h\to 0^+}\frac{f(0+h)-f(0)}{h} = \lim_{h\to 0^+}\frac{(2h+1)-(1)}{h} = \lim_{h\to 0^-}\frac{2h}{h} = 2$$

Because these two limits are not the same, the derivative *does not* exist at $x = 0$.

2. Evaluate the derivative using the definition:

$$f'(0) = \lim_{h\to 0^-}\frac{f(0+h)-f(0)}{h} = \lim_{h\to 0^-}\frac{\left(h^2\right)-(0)}{h} = \lim_{h\to 0^-}\frac{h^2}{h} = 0$$

$$f'(0) = \lim_{h \to 0^+} \frac{f(0+h)-f(0)}{h} = \lim_{h \to 0^+} \frac{(h^3)-(0)}{h} = \lim_{h \to 0^-} \frac{h^3}{h} = 0$$

Because the two limits are the same, the derivative *does* exist at $x = 0$.

Lesson 7-5 Review

1. The point is (2, 10) and $f'(x) = 2x + 3$, so the slope is $f'(2) = 7$.
   The equation of the tangent line is $y - 10 = 7(x - 2)$, or $y = 7x - 4$.

2. The point is (2, 2) and $f'(x) = \frac{1}{2\sqrt{x+2}}$, so the slope is $f'(2) = \frac{1}{4}$.

   The equation of the tangent line is $y - 2 = \frac{1}{4}(x - 2)$, or $y = \frac{1}{4}x + \frac{3}{2}$.

3. The point is (−3, 1) and $f'(x) = e^{x+3}$, so the slope is $f'(-3) = 1$.
   The equation of the tangent line is $y - 1 = 1(x + 3)$, or $y = x + 4$.

4. The point is (1, 1) and $f'(x) = -\frac{1}{x^2}$, so the slope is $f'(1) = -1$.
   The equation of the tangent line is $y - 1 = -1(x - 1)$, or $y = -x + 2$.

5. The point is $\left(\frac{\pi}{2}, 0\right)$ and $f'(x) = -\sin x$, so the slope is $f'\left(\frac{\pi}{2}\right) = -1$.

   The equation of the tangent line is $y - 0 = -1\left(x - \frac{\pi}{2}\right)$, or $y = -x + \frac{\pi}{2}$.

Rules for Differentiation

Evaluating the derivative using the definition can be time consuming. Although we have some techniques to help us evaluate limits, if the only way to take the derivative of a function was to evaluate $\lim_{h \to 0} \frac{f(x+h)-f(x)}{h}$ algebraically, calculus would not be as popular as it is. In this chapter we will discuss some short cuts for taking the derivative of a function. Two important rules that we will discuss are the product rule and the quotient rule.

Lesson 8-1: Sums and Differences

There are a few functions whose derivative is relatively easy to determine just by using the *interpretation* of the derivative. The derivative of a function represents the slope of the line tangent to the graph of the function. We will first consider the constant function: $f(x) = a$. The graph of this function is the horizontal line $y = a$. Lines that are tangent to this line will also be horizontal lines, and the slope of a horizontal line is 0. Therefore, the derivative of a constant is 0. This is true for *any* constant.

The next function to analyze is $f(x) = mx$. The graph of this function is a line with slope m. Lines tangent to this function will also have slope m, and the *derivative* of the function $f(x) = mx$ is m.

The rule for taking the derivative of the sum of two functions is based on the fact that the limit of a sum is the sum of the limits:

The derivative of the sum is the sum of the derivatives.

This means that if $f(x) = g(x) + h(x)$, finding $f'(x)$ can be done by finding $g'(x)$ and $h'(x)$ individually, and then adding the results. A similar relationship holds for the derivative of the difference between two functions: $(f(x) - g(x))' = f'(x) - g'(x)$.

We can use these two results and the first property of the derivative (the derivative of the sum is the sum of the derivative) to find the derivative of any *linear* function.

Example 1

Find the derivative of the following functions:

a. $f(x) = 3x + 4$

b. $f(x) = \pi x + e^2$

c. $f(x) = \left(\sin\frac{\pi}{4}\right)x + \cosh 2$

Solution: All of these functions are linear, with slope equal to the coefficient in front of x. The derivative of these functions will be their respective slopes:

a. $f'(x) = 3$

b. $f'(x) = \pi$

c. $f'(x) = \left(\sin\frac{\pi}{4}\right)$

There is another property of limits that will help us find the derivative of more complicated functions. The second property of limits states that $\lim\limits_{x\to c}\left[k \cdot f(x)\right] = k \cdot \lim\limits_{x\to c} f(x)$. In other words, the constant can be brought outside of the limit, because a constant does not care what x approaches. When taking the derivative of the product of a constant and a function, the constant waits patiently while the function is being differentiated. The constant does not go away, but it also does not change. This rule can be written as $(kf(x))' = kf'(x)$.

Lesson 8-1 Review

Find the derivative of the following functions:

1. $f(x) = 4x + \ln 5$

2. $f(x) = -2x - 4$

3. $f(x) = \frac{1}{3}x + 5$

Lesson 8-2: The Power Rule

The power rule enables us to find the derivative of any power function. Using the fact that the derivative of the sum is the sum of the derivative, you will then be able to find the derivative of any polynomial. The power rule will also play a role in establishing the chain rule. The power rule is as follows:

If $f(x) = x^n$ where a and n are constants, then $f'(x) = nx^{n-1}$.

We can practice using this first short cut for taking the derivative.

Example 1

Differentiate the following functions:

a. $f(x) = x^4$

b. $g(x) = \sqrt{x}$

c. $h(x) = x^{-3}$

d. $f(x) = \frac{1}{x^2}$

Solution: Apply the power rule: If $f(x) = x^n$, then $f'(x) = nx^{n-1}$. If the function is not explicitly written in the form $f(x) = x^n$, rewrite the function so that it is in this form.

a. $f(x) = x^4$: $f'(x) = 4x^{4-1} = 4x^3$

b. $g(x) = \sqrt{x}$: Rewrite this function so that the exponent is a fraction.

Then use the power rule: $g(x) = x^{1/2}$, and $g'(x) = \frac{1}{2}x^{1/2-1} = \frac{1}{2}x^{-1/2} = \frac{1}{2\sqrt{x}}$

c. $h(x) = x^{-3}$: $h'(x) = -3x^{-3-1} = -3x^{-4}$

d. $f(x) = \frac{1}{x^2}$: Rewrite this function so that the exponent is a negative integer. Then use the power rule: $f(x) = x^{-2}$ and $f'(x) = -2x^{-2-1} = -2x^{-3}$.

It is important that you avoid the traps set by disguised power functions. Rewrite the functions so that you can use the power rule directly. It may be tempting to save time by trying to work with functions such as $f(x)=\frac{1}{x^2}$ or $g(x)=\sqrt{x}$ without rewriting the function. If you try to skip steps, the result could be costly. By taking a little extra time, you can be sure that you are taking the derivative correctly.

Now we are ready to take the derivative of more complicated functions. We can differentiate any polynomial, any sum of power functions, or any constant times a power function.

Example 2

Differentiate the following functions:

a. $f(x)=4x^3+6\sqrt{x}$

b. $h(x) = 5x^{-3} - 3x^2 + 2$

c. $f(x)=\frac{1}{x^2}-\frac{1}{x^5}$

Solution: Apply the power rule and the properties of the derivative. Rewrite any power functions that are not in their proper format.

a. $f(x)=4x^3+6\sqrt{x}=4x^3+6x^{1/2}$:

$f'(x)=4\left(3x^{3-1}\right)+6\left(\frac{1}{2}x^{1/2-1}\right)=12x^2+3x^{-1/2}$

b. $h(x) = 5x^{-3} - 3x^2 + 2$: $h'(x) = 5(-3x^{-3-1}) - 3(2x^{2-1}) + 0 = -15x^{-4} - 6x$

c. $f(x)=\frac{2}{x^2}-\frac{3}{x^5}=2x^{-2}-3x^{-5}$:

$f'(x) = 2(-2x^{-2-1}) - 3(-5x^{-5-1}) = -4x^{-3} + 15x^{-6}$

Now that we can find the derivative of a variety of functions, we can practice using the derivative to find the equations of tangent lines. To find the equation of a tangent line, we need a point and a slope. The point can be found by evaluating the *function* at the specific x-coordinate. The slope can be found by evaluating the *derivative* of the function at the specific x-coordinate.

Example 3

Find the equation of the line tangent to the following functions at the point indicated. Write your equations in slope-intercept form.

a. $f(x)=3x^2+\sqrt{x}$ at $x=1$

b. $g(x)=\frac{4}{x}$ at $x=2$

Solution: Find the point and the slope, and then use the point-slope formula:

a. $f(x)=3x^2+\sqrt{x}$ at $x=1$: The point is $(1,4)$. To find the slope, evaluate the derivative at $x = 1$: $f'(x)=6x+\frac{1}{2}x^{-1/2}$, and $f'(1)=\frac{13}{2}$. Use the point-slope formula:

$$y-4=\tfrac{13}{2}(x-1)$$
$$y=\tfrac{13}{2}x-\tfrac{5}{2}$$

The graphs of $f(x)=3x^2+\sqrt{x}$ and of $y=\frac{13}{2}x-\frac{5}{2}$ are shown in Figure 8.1.

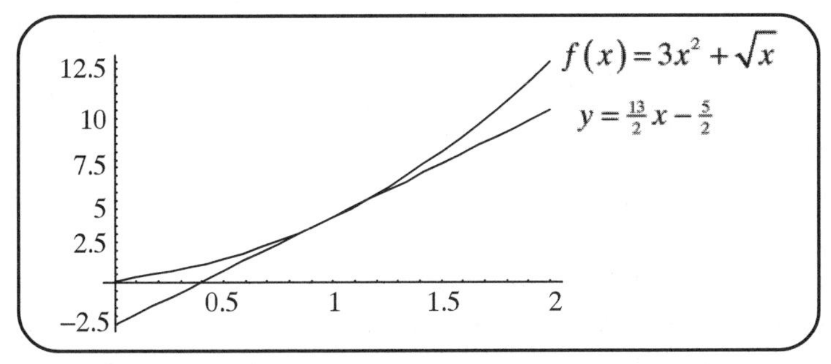

Figure 8.1.

b. $g(x)=\frac{4}{x}$ at $x=2$: The point is (2, 2). To find the slope, evaluate the derivative at $x = 2$: $g'(x) = -4x^{-2}$, and $g'(2) = -1$. Use the point-slope formula:

$$y-2=-1(x-2)$$
$$y=-x+4$$

The graphs of $g(x)=\frac{4}{x}$ and $y = -x + 4$ are shown in Figure 8.2.

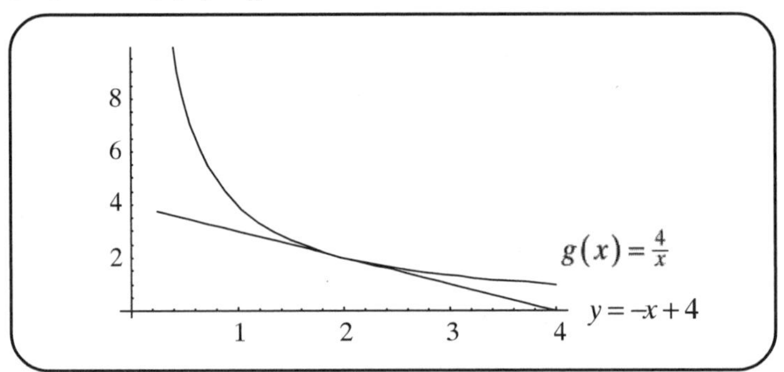

Figure 8.2.

Lesson 8-2 Review

1. Differentiate the following functions:

   a. $f(x)=\sqrt[3]{x}+\frac{2}{x}$

   b. $g(x)=8x-\frac{3}{x^2}$

2. Find the equation of the line tangent to the following functions at the point indicated. Write your equations in slope-intercept form.

   a. $f(x) = x^4 + 2x + 1$ at $x = 1$

   b. $g(x)=\frac{4}{x}-2x$ at $x=2$

Lesson 8-3: The Product Rule

We can now turn our attention to differentiating more complicated functions. The derivative of a function measures how a change in the independent variable affects, or changes, the dependent variable. The difference quotient is the ratio of these changes: $\frac{\Delta f}{\Delta x}$, and the derivative is $\lim_{\Delta x \to 0} \frac{\Delta f}{\Delta x}$. In this lesson, we will examine how a product of functions changes as the independent variable changes.

As the independent variable of a product of two functions changes, *both* of the functions involved in the product will change. Too much change can result in a sensory overload, so we only allow one function to change at a time. The product rule is as follows:

$$(f(x) \cdot g(x))' = f'(x) \cdot g(x) + f(x) \cdot g'(x)$$

In other words, when taking the derivative of the product of two functions, take the derivative of the first function (changed) and multiply it by the second function (unchanged). Then take the derivative of the second function (changed) and multiply it by the first function (unchanged). Add the two results together. In effect, the derivative of the product of two functions is the derivative of the first times the second plus the first one left alone times the derivative of the second. Not only is it important that you learn what the product rule says, but how to use it correctly.

Example 1

Differentiate the following functions. Do not simplify.

a. $f(x) = (4x^2 + 1)(6x - 2)$

b. $g(x) = \left(\sqrt{x} + \frac{1}{x}\right)(x-2)$

Solution: Apply the product rule:

a. $f(x) = (4x^2 + 1)(6x - 2)$:

$f'(x) = (4x^2 + 1)'(6x - 2) + (4x^2 + 1)(6x - 2)'$

$f'(x) = (8x)(6x - 2) + (4x^2 + 1)(6)$

b. $g(x) = \left(\sqrt{x} + \frac{1}{x}\right)(x-2)$:

$$g'(x) = \left(\sqrt{x} + \frac{1}{x}\right)'(x-2) + \left(\sqrt{x} + \frac{1}{x}\right)(x-2)'$$

$$g'(x) = \left(x^{1/2} + x^{-1}\right)'(x-2) + \left(\sqrt{x} + \frac{1}{x}\right)(x-2)'$$

$$g'(x) = \left(\frac{1}{2}x^{-1/2} - x^{-2}\right)(x-2) + \left(\sqrt{x} + \frac{1}{x}\right)(1)$$

Pay attention to the notation. When I apply the product rule, I put *primes* on the pieces that need to be differentiated. The product rule says to take the derivative of the first times the second plus the first one left alone times the derivative of the second. Notice that the first step in differentiating the product of two functions is to figure out which piece needs to be differentiated at each stage. Take an extra step to apply the product rule correctly and then take the derivatives of each of the individual functions. Use a systematic approach with these problems. If you establish a routine early and are careful at each step along the way, differentiating products of functions should be no problem.

In the previous examples, I specifically indicated that you should not simplify. Right now, the focus is on the mechanics of taking the derivative. Multiplying polynomials together and collecting terms is a skill that has been practiced enough in algebra; you should be able to simplify these expressions when necessary.

We can now find equations of lines tangent to more complicated functions at a given point. Tangent lines are important in calculus, and we will discuss them in more detail in Chapter 11. For now, we will focus on the *mechanics* of finding equations of tangent lines.

Example 2

Find the equation of the line tangent to the given functions at the indicated point.

a. $f(x)=\left(3x^3+\frac{2}{\sqrt{x}}\right)\left(6x^2-2x\right)$ at $x=1$

b. $h(x)=\left(\frac{2}{x^3}-\frac{3}{x^4}\right)\left(2x^4-7x^2+5\right)$ at $x=1$

Solution: Evaluate the function at the given x-coordinate to find the point, and find the slope by evaluating the *derivative* of the function at the given x-coordinate. Then use the point-slope formula. To verify our answers, both the function and the tangent line will be graphed. The tangent lines should just touch the graph of the functions at the given point.

a. $f(x)=\left(3x^3+\frac{2}{\sqrt{x}}\right)\left(6x^2-2x\right)$ at $x=1$: The point is (1, 20).

The derivative is:

$$f'(x)=\left(3x^3+\frac{2}{\sqrt{x}}\right)'\left(6x^2-2x\right)+\left(3x^3+\frac{2}{\sqrt{x}}\right)\left(6x^2-2x\right)'$$

$$f'(x)=\left(9x^2+2\left(-\tfrac{1}{2}x^{-3/2}\right)\right)\left(6x^2-2x\right)+\left(3x^3+\tfrac{2}{\sqrt{x}}\right)(12x-2)$$

$$f'(1)=82$$

Finally, use the point-slope formula:

$$y-20=82(x-1)$$
$$y=82x-62$$

The graphs of $f(x)=\left(3x^3+\frac{2}{\sqrt{x}}\right)\left(6x^2-2x\right)$ at $x=1$ and $y=82x-62$ are shown in Figure 8.3.

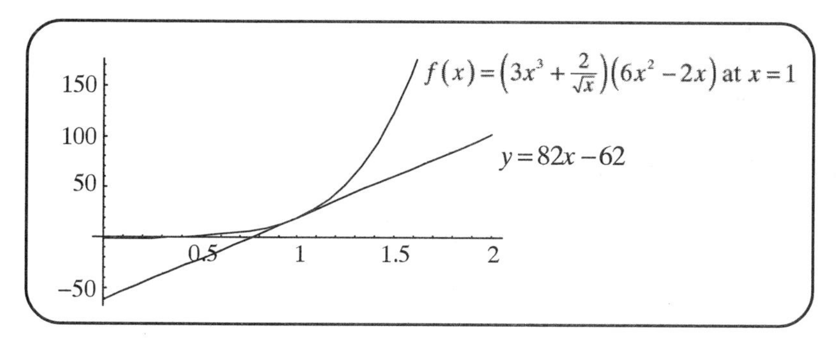

Figure 8.3

b. $h(x)=\left(\frac{2}{x^3}-\frac{3}{x^4}\right)\left(2x^4-7x^2+5\right)$ at $x=1$: The point is $(1, 0)$. To find the slope of the tangent line, evaluate $h'(1)$:

$$h'(x)=\left(\tfrac{2}{x^3}-\tfrac{3}{x^4}\right)'\left(2x^4-7x^2+5\right)+\left(\tfrac{2}{x^3}-\tfrac{3}{x^4}\right)\left(2x^4-7x^2+5\right)'$$

$$h'(x)=\left(-6x^{-4}+12x^{-5}\right)\left(2x^4-7x^2+5\right)+\left(\tfrac{2}{x^3}-\tfrac{3}{x^4}\right)\left(8x^3-14x\right)$$

$$h'(1)=(-6+12)(2-7+5)+(2-3)(8-14)=6$$

Finally, use the point-slope formula:

$$y-0=6(x-1)$$
$$y=6x-6$$

The graphs of $h(x)=\left(\frac{2}{x^3}-\frac{3}{x^4}\right)(2x^4-7x^2+5)$ and $y = 6x - 6$ are shown in Figure 8.4.

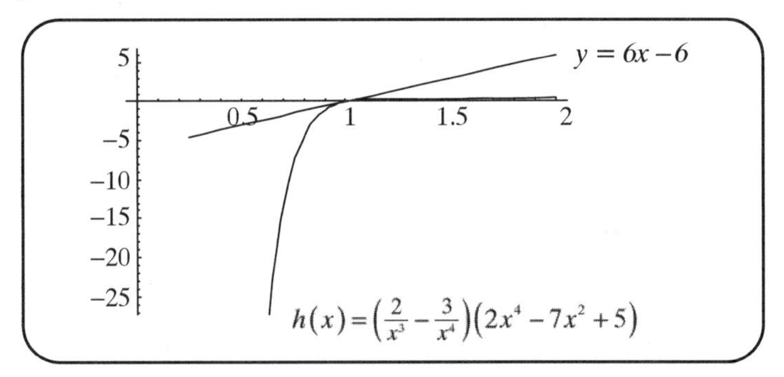

Figure 8.4

Lesson 8-3 Review

1. Differentiate the following functions:
   a. $f(x) = (4x^6 - 3x^2 + 2)(x^9 - 8x + 1)$
   b. $g(x)=(\sqrt{x}-5x)(2x^4-7x^2+5)$

2. Find the equation of the line tangent to the graph of the following functions at the indicated point:
   a. $f(x)=\left(x+\frac{3}{x}\right)(x^4-2x+1)$, at $x=1$
   b. $g(x)=(\sqrt{x}-2)(\sqrt{x}+5)$ at $x=1$

Lesson 8-4: The Quotient Rule

Another important differentiating rule is the quotient rule. The quotient rule will enable us to evaluate the derivative of the quotient, or ratio, of two functions. It is similar to the product rule, but there are enough differences that you need to be careful when applying it. The derivative of the quotient of two functions is given by the following formula:

$$\left(\frac{f(x)}{g(x)}\right)'=\frac{f'(x)g(x)-f(x)g'(x)}{(g(x))^2}$$

Notice the similarities between the quotient rule and the product rule. The numerator of the derivative involves taking the derivative of the numerator and the derivative of the denominator of the function one at a time, just as we did using the product rule. Instead of the results being added, they are subtracted. This rule actually comes from the product rule, though we can also derive it using the chain rule, which we will discuss in Chapter 10. Right now, we will focus on applying the quotient rule and reinforcing our tangent line skills.

Example 1

Differentiate the following functions. Do not simplify.

a. $f(x)=\frac{(4x^2+1)}{(6x-2)}$

b. $g(x)=\frac{\left(\sqrt{x}+\frac{1}{x}\right)}{(x^2+1)}$

Solution: Apply the quotient rule.

a. $f(x)=\frac{(4x^2+1)}{(6x-2)}$: $f'(x)=\frac{(4x^2+1)'(6x-2)-(4x^2+1)(6x-2)'}{(6x-2)^2}$

$$f'(x)=\frac{(8x)(6x-2)-(4x^2+1)(6)}{(6x-2)^2}$$

b. $g(x)=\frac{\left(\sqrt{x}+\frac{1}{x}\right)}{(x^2+1)}$: $g'(x)=\frac{\left(\sqrt{x}+\frac{1}{x}\right)'(x^2+1)-\left(\sqrt{x}+\frac{1}{x}\right)(x^2+1)'}{(x^2+1)^2}$

$$g'(x)=\frac{\left(\frac{1}{2}x^{-1/2}-x^{-2}\right)(x^2+1)-\left(\sqrt{x}+\frac{1}{x}\right)(2x)}{(x^2+1)^2}$$

We can find equations of tangent lines and work out other types of applications.

Example 2

Where does the graph of the function $f(x)=\frac{x}{x^2+1}$ have a horizontal tangent line?

Solution: The slope of the tangent line can be found by taking the derivative of the function. The slope of a horizontal line is 0. In order to answer this question, we need to find where the derivative is equal to 0. First, find the derivative using the quotient rule:

$$f'(x)=\frac{(x)'(x^2+1)-(x)(x^2+1)'}{(x^2+1)^2}$$

$$f'(x)=\frac{(1)(x^2+1)-(x)(2x)}{(x^2+1)^2}$$

Now, set the derivative equal to 0 and solve for x:

$$\frac{(1)(x^2+1)-(x)(2x)}{(x^2+1)^2}=0$$

$$\frac{(x^2+1)-2x^2}{(x^2+1)^2}=0$$

$$\frac{-x^2+1}{(x^2+1)^2}=0$$

Now, $f'(x)=0$ when the numerator is equal to 0:

$$-x^2+1=0$$

$$x^2=1$$

$$x^2=\pm 1$$

The graph of the function $f(x)=\frac{x}{x^2+1}$ has horizontal tangent lines at $x=\pm 1$.

The graph of $f(x)=\frac{x}{x^2+1}$ is shown in Figure 8.5. This function is an odd function. We can see this algebraically:

$$f(-x)=\frac{(-x)}{(-x)^2+1}=-\frac{x}{x^2+1}=-f(x)$$

We can also see this because the numerator is an odd function and the denominator is an even function, and an odd function divided by an even function will be an odd function. Either way, we expect the graph to be symmetric about the origin, and we expect the tangent lines to be horizontal at $x = \pm 1$.

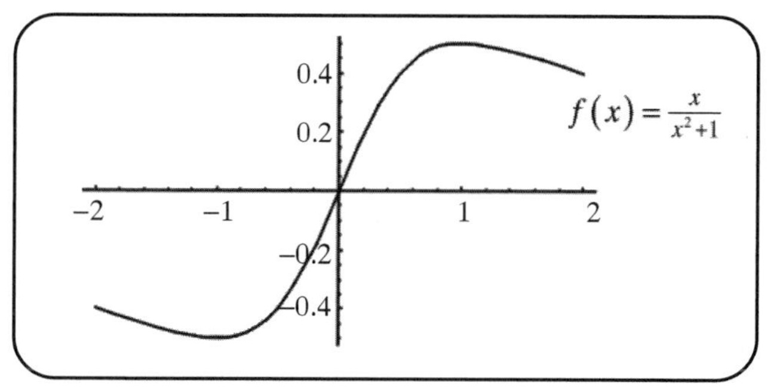

Figure 8.5

We can differentiate some functions by using either the power rule or the quotient rule. Let's differentiate the function $f(x)=\frac{4}{x}$ using both techniques:

Power Rule	Quotient Rule
$f(x)=4x^{-1}$	$f(x)=\frac{4}{x}$
$f'(x)=-4x^{-2}$	$f'(x)=\frac{(4)'(x)-(4)(x)'}{(x)^2}$
	$f'(x)=\frac{(0)(x)-(4)(1)}{(x)^2}$
	$f'(x)=-\frac{4}{x^2}=-4x^{-2}$

Either way, we get the same result. It is natural to treat $f(x)=\frac{4}{x}$ as a quotient and differentiate it using the quotient rule, but using the power rule involves fewer steps and less simplification. Regardless of the rule you use to take a derivative, if you apply the rule correctly, and carefully differentiate each piece of the function at the right time, you cannot go wrong.

Lesson 8-4 Review

1. Differentiate the following functions:

   a. $f(x)=\dfrac{\left(\frac{2}{x}-4x\right)}{\left(2x^4-7x^2+5\right)}$

   b. $g(x)=\dfrac{x^4+7x+1}{\sqrt{x}+10}$

2. Find the equation of the line tangent to the following functions at the indicated point:

   a. $f(x)=\dfrac{4x+2}{x^2+1}$ at $x=0$

   b. $g(x)=\dfrac{2x-1}{3x+2}$ at $x=-1$

3. Where does the graph of the function $f(x)=\dfrac{(2x+1)}{(x^2+4)}$ have a horizontal tangent line?

4. Where does the graph of the function $f(x)=\dfrac{(2x+1)}{(x+4)}$ have a tangent line with slope equal to 1?

Answer Key

Lesson 8-1 Review

1. $f'(x)=4$
2. $f'(x)=-2$
3. $f'(x)=\frac{1}{3}$

Lesson 8-2 Review

1. a. $f(x)=\sqrt[3]{x}+\frac{2}{x}=x^{1/3}+2x^{-1},\ f'(x)=\frac{1}{3}x^{-2/3}-2x^{-2}$

   b. $g(x)=8x-\frac{3}{x^2}=8x-3x^{-2},\ g'(x)=8+6x^{-3}$

2. a. $f(x) = x^4 + 2x + 1$ at $x = 1$:
The point is $(1, 4)$, $f'(x) = 4x^3 + 2$, so the slope is $f'(1) = 6$.
The equation of the tangent line is $y - 4 = 6(x - 1)$, or $y = 6x - 2$.

b. $g(x) = \frac{4}{x} - 2x$ at $x = 2$:
The point is $(2, -2)$, $g'(x) = -4x^{-2} - 2$, so the slope is $g'(2) = -3$.
The equation of the tangent line is $y + 2 = -3(x - 2)$, or $y = -3x + 4$.

Lesson 8-3 Review

1. a. $f(x) = (4x^6 - 3x^2 + 2)(x^9 - 8x + 1)$
$f'(x) = (4x^6 - 3x^2 + 2)'(x^9 - 8x + 1) + (4x^6 - 3x^2 + 2)(x^9 - 8x + 1)'$
$f'(x) = (24x^5 - 6x)(x^9 - 8x + 1) + (4x^6 - 3x^2 + 2)(9x^8 - 8)$

b. $g(x) = \left(\sqrt{x} - 5x\right)\left(2x^4 - 7x^2 + 5\right)$

$g'(x) = \left(\sqrt{x} - 5x\right)'\left(2x^4 - 7x^2 + 5\right) + \left(\sqrt{x} - 5x\right)\left(2x^4 - 7x^2 + 5\right)'$

$g'(x) = \left(\frac{1}{2}x^{-1/2} - 5\right)\left(2x^4 - 7x^2 + 5\right) + \left(\sqrt{x} - 5x\right)\left(8x^3 - 14x\right)$

2. a. $f(x) = \left(x + \frac{3}{x}\right)\left(x^4 - 2x + 1\right)$ at $x = 1$:

The point is $(1, 0)$, and $f'(x) = \left(1 - 3x^{-2}\right)\left(x^4 - 2x + 1\right) + \left(x + \frac{3}{x}\right)\left(4x^3 - 2\right)$,
so the slope is $f'(1) = 8$.
The equation of the tangent line is $y - 0 = 8(x - 1)$, or $y = 8x - 8$.

b. $g(x) = \left(\sqrt{x} - 2\right)\left(\sqrt{x} + 5\right)$ at $x = 1$:

The point is $(1, -6)$, and $g'(x) = \left(\frac{1}{2}x^{-1/2}\right)\left(\sqrt{x} + 5\right) + \left(\sqrt{x} - 2\right)\left(\frac{1}{2}x^{-1/2}\right)$,

so the slope is $g'(1) = \frac{5}{2}$.

The equation of the tangent line is $y + 6 = \frac{5}{2}(x - 1)$, or $y = \frac{5}{2}x - \frac{17}{2}$.

Lesson 8-4 Review

1. a. $f(x) = \frac{\left(\frac{2}{x} - 4x\right)}{\left(2x^4 - 7x^2 + 5\right)}$: $f'(x) = \frac{\left(\frac{2}{x} - 4x\right)'\left(2x^4 - 7x^2 + 5\right) - \left(\frac{2}{x} - 4x\right)\left(2x^4 - 7x^2 + 5\right)'}{\left(2x^4 - 7x^2 + 5\right)^2}$

$$f'(x)=\frac{\left(-2x^{-2}-4\right)\left(2x^4-7x^2+5\right)-\left(\frac{2}{x}-4x\right)\left(8x^3-14x\right)}{\left(2x^4-7x^2+5\right)^2}$$

b. $g(x)=\frac{x^4+7x+1}{\sqrt{x}+10}$: $g'(x)=\frac{\left(x^4+7x+1\right)'\left(\sqrt{x}+10\right)-\left(x^4+7x+1\right)\left(\sqrt{x}+10\right)'}{\left(\sqrt{x}+10\right)^2}$

$$g'(x)=\frac{\left(4x^3+7\right)\left(\sqrt{x}+10\right)-\left(x^4+7x+1\right)\left(\frac{1}{2}x^{-1/2}\right)}{\left(\sqrt{x}+10\right)^2}$$

2. a. $f(x)=\frac{4x+2}{x^2+1}$ at $x=0$:

The point is (0, 2) and $f'(x)=\frac{(4)\left(x^2+1\right)-(4x+2)(2x)}{\left(x^2+1\right)^2}$, so the slope is $f'(0)=4$.

The equation of the tangent line is $y-2=4(x-0)$, or $y=4x+2$.

b. $g(x)=\frac{2x-1}{3x+2}$ at $x=-1$:

The point is (−1, 3) and $g'(x)=\frac{(2)(3x+2)-(2x-1)(3)}{(3x+2)^2}$, so the slope is $g'(0)=7$.

The equation of the tangent line is $y-3=7(x+1)$, or $y=7x+10$.

3. Take the derivative, set the numerator of the derivative equal to 0, and solve for x:

$$f'(x)=\frac{(2)\left(x^2+4\right)-(2x+1)(2x)}{\left(x^2+4\right)^2}=0$$

$$(2)(x^2+4)-(2x+1)(2x)=0$$
$$2x^2+8-4x^2-2x=0$$
$$-2x^2-2x+8=0$$

$$x=\frac{2\pm\sqrt{4-4(-2)(8)}}{2(-2)}$$

$$x=\frac{2\pm\sqrt{68}}{-4}=\frac{-1\pm\sqrt{17}}{2}$$

The graph of $f(x)=\frac{(2x+1)}{\left(x^2+4\right)}$ has a horizontal tangent line at $x=\frac{-1+\sqrt{17}}{2}$ and

$x=\frac{-1-\sqrt{17}}{2}$.

4. Take the derivative, set it equal to 1, and solve for x:

$$f'(x)=\frac{(2)(x+4)-(2x+1)(1)}{(x+4)^2}=1$$

$$\frac{2x^2+8-4x^2-2x}{(x+4)^2}=1$$

$$-2x^2-2x+8=(x+4)^2$$
$$-2x^2-2x+8=x^2+8x+16$$
$$3x^2+10x+8=0$$
$$(3x+4)(x+2)=0$$

The graph of $f(x)=\frac{2x+1}{x+4}$ has a tangent line with slope equal to 1 at $x=-2$ and $x=-\frac{4}{3}$.

Derivatives of Exponential and Trigonometric Functions

Using the techniques discussed in Chapter 8, we can differentiate sums, differences, products, and quotients of power functions. In this chapter we will develop rules for differentiating exponential and trigonometric functions. We will also introduce a new set of functions, called the hyperbolic functions, which are defined in terms of exponential functions. These functions are used in physics and engineering, and we will analyze these new functions using calculus. When we are through with this chapter, you will be able to take the derivative of a variety of interesting functions.

Lesson 9-1 Exponential Functions

We will now turn our attention to simple exponential functions. In general, the derivative of an exponential function $f(x) = a^x$ is given by $f'(x) = a^x \ln a$. A special case of the rule for differentiating exponential functions occurs when the base is e. The derivative of the function $f(x) = e^x$ is $f'(x) = e^x \ln e = e^x$. The derivative of the function $f(x) = e^x$ is just itself. This leads us to another way to define the number e.

The number e can be defined as $\lim_{n\to\infty}\left(1+\frac{1}{n}\right)^n$. This was interpreted as being the amount of money obtained after 1 year if $1 was invested at 100% interest compounded continuously. Another way to obtain the same number e is to determine the base of the exponent, a, that satisfies the equality $\lim_{h\to 0}\frac{a^h-1}{h}=1$. Surprisingly enough, the base of the exponent a that satisfies the equation $\lim_{h\to 0}\frac{a^h-1}{h}=1$ is e. There are many ways to generate e from a limit, and we will see several examples of this in Chapter 12.

Now that we can differentiate simple exponential functions, we can differentiate more complicated functions created from combinations of exponential functions and power functions. We will practice using the product and quotient rules with exponential and power functions.

Example 1

Differentiate the following functions. Do not simplify.

a. $f(x) = e^x(x^2 + 2x + 1)$

b. $g(x)=\frac{e^x}{(x^2+1)}$

Solution: Apply the product or quotient rule:

a. $f(x) = e^x(x^2 + 2x + 1)$: $f'(x) = (e^x)'(x^2 + 2x + 1) + (e^x)(x^2 + 2x + 1)'$

$$f'(x) = (e^x)(x^2 + 2x + 1) + (e^x)(2x + 2)$$

b. $g(x)=\frac{e^x}{(x^2+1)}$: $g'(x)=\frac{(e^x)'(x^2+1)-(e^x)(x^2+1)'}{(x^2+1)^2}$

$$g'(x)=\frac{(e^x)(x^2+1)-(e^x)(2x)}{(x^2+1)^2}$$

Example 2

Where does the graph of the function $f(x) = xe^x$ have a horizontal tangent line?

Solution: If the tangent line is horizontal, its slope is 0. Therefore, we need to determine where the derivative of $f(x) = xe^x$ is equal to 0. First, take the derivative of the function: $f'(x) = (x)'e^x + (x)(e^x)'$

$$f'(x) = (1)e^x + (x)(e^x)$$

Next, set the derivative equal to 0 and solve for x:

$$f'(x) = (1)e^x + (x)(e^x) = 0$$

Factor out e^x from both terms: $(e^x)(1+x)=0$

Set each factor equal to 0. Remember that $e^x \neq 0$ for all real numbers x.

$$(1+x)=0$$
$$x=-1$$

The graph of $f(x) = xe^x$ is shown in Figure 9.1. Notice that at $x = -1$, the tangent line is horizontal.

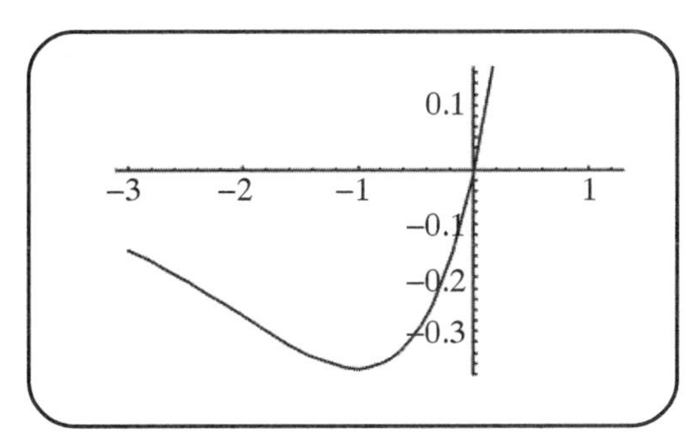

Figure 9.1.

Example 3

Find the equation of the line tangent to the given functions at the indicated point.

a. $f(x) = e^x(6x^2 - 2x + 3)$ at $x = 0$

b. $g(x)=\frac{4e^x}{x^2+1}$ at $x=0$

Solution: Find the point (by evaluating the function at the indicated value of x), find the slope (by evaluating the derivative of the function at the indicated value of x), and then use the point-slope formula:

a. $f(x) = e^x(6x^2 - 2x + 3)$ at $x = 0$: The point is $(0, 3)$.

The derivative is: $f'(x) = (e^x)'(6x^2 - 2x + 3) + (e^x)(6x^2 - 2x + 3)'$

$$f'(x) = (e^x)(6x^2 - 2x + 3) + (e^x)(12x - 2)$$
$$f'(0) = (1)(3) + (1)(-2) = 1$$

Now use the point-slope formula and simplify:

$$y-3=1(x-0)$$
$$y=x+3$$

The graphs of $f(x)=e^x(6x^2-2x+3)$ and $y=x+3$ are shown in Figure 9.2.

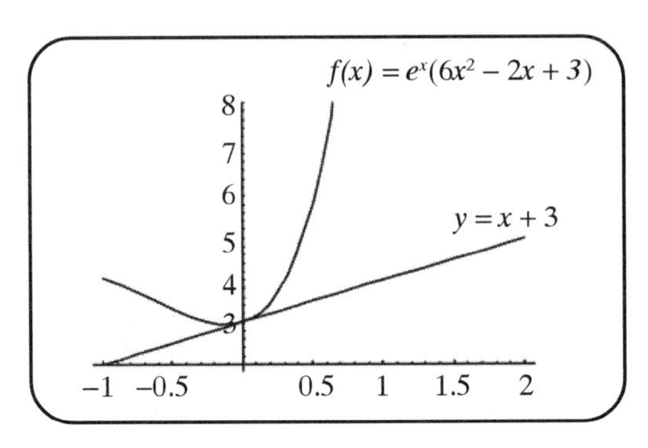

Figure 9.2.

b. $g(x)=\frac{4e^x}{x^2+2}$ at $x=0$: The point is (0, 2).

The derivative is: $$g'(x)=\frac{(4e^x)'(x^2+2)-(4e^x)(x^2+2)'}{(x^2+2)^2}$$

$$g'(x)=\frac{(4e^x)(x^2+2)-(4e^x)(2x)}{(x^2+2)^2}$$

$$g'(x)=\frac{(4)(2)-(4)(0)}{(2)^2}=\frac{8}{4}=2$$

Now use the point-slope formula and simplify:

$$y-2=2(x-0)$$
$$y=2x+2$$

The graphs of $g(x)=\frac{4e^x}{x^2+2}$ and $y=2x+2$ are shown in Figure 9.3 on page 157.

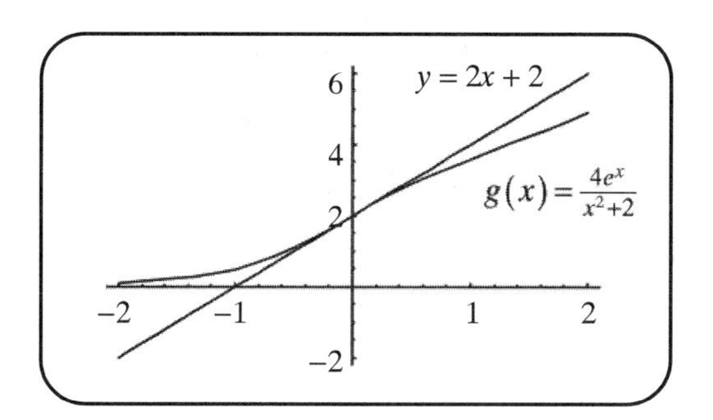

Figure 9.3.

Lesson 9-1 Review

1. Differentiate the following functions:

   a. $h(x)=\frac{(3^x-2^x+e^2)}{(2x^4-7x^2+5)}$ b. $f(x)=\frac{x}{e^x+1}$

2. Find the equation of the line tangent to the following functions at the indicated point:

   a. $f(x) = (x^2 + 1)e^x$ at $x = 0$ b. $g(x)=\frac{xe^x}{x+1}$ at $x=0$

3. Where does the graph of the function $f(x) = x^2e^x$ have a horizontal tangent line?

Lesson 9-2: Hyperbolic Functions

The hyperbolic functions are linear combinations of the functions e^x and e^{-x}. The **hyperbolic sine** function, abbreviated $\sinh x$ (and pronounced "cinch"), is defined as:

$$\sinh x = \frac{e^x - e^{-x}}{2}$$

The **hyperbolic cosine** function, abbreviated $\cosh x$ (and pronounced so that it rhymes with "gosh"), is defined as:

$$\cosh x = \frac{e^x + e^{-x}}{2}$$

Before we differentiate these functions, we should take a minute to explore some of their basic properties. These functions are defined in a symmetrical way, and they may inherit some symmetry as a result. First, let's find the y-intercepts of these functions:

$$\sinh(0)=\frac{e^0-e^0}{2}=0 \text{ and } \cosh(0)=\frac{e^0+e^0}{2}=1$$

The hyperbolic sine function passes through the origin, and the y-intercept of the hyperbolic cosine is (0, 1).
The hyperbolic sine function is an odd function:

$$\sinh(-x)=\frac{e^{-x}-e^{-(-x)}}{2}=\frac{e^{-x}-e^{x}}{2}=-\left(\frac{e^{x}-e^{-x}}{2}\right)=-\sinh(x)$$

The hyperbolic cosine function is an even function:

$$\cosh(-x)=\frac{e^{-x}+e^{-(-x)}}{2}=\frac{e^{-x}+e^{x}}{2}=\frac{e^{x}+e^{-x}}{2}=\cosh(x)$$

The graph of the hyperbolic sine function will be symmetric about the origin, and the graph of the hyperbolic cosine function will be symmetric about the y-axis. The graph of the hyperbolic cosine function can be used to describe the shape of a chain that is suspended between two points that have the same height, and that shape is called a catenary curve.

From the equation $\cosh x=\frac{e^x+e^{-x}}{2}$, we see that $\cosh x > 0$ for all x. This function will never cross the x-axis. From the equation $\sinh x=\frac{e^x-e^{-x}}{2}$, we see that the only place where $\sinh x$ crosses the x-axis is at $x = 0$.

In addition to the symmetry of the hyperbolic functions, it is worthwhile to examine the asymptotic behavior of these functions. As x gets large, e^x gets large and e^{-x} becomes negligible. As $x\to\infty$, both sinh (x) and cosh (x) behave like e^x. We can write that as $x\to\infty$, sinh (x) ~ e^x and cosh (x) ~ e^x. On the other hand, if x is large in magnitude but is negative, then e^x becomes negligible and e^{-x} is large and dominates. As $x\to-\infty$, sinh (x) behaves as $-e^{-x}$ and cosh (x) behaves as e^{-x}. We can write that as $x\to-\infty$, sinh (x) ~ $-e^{-x}$ and cosh (x) ~ e^{-x}. Neither of these functions has a horizontal asymptote; their asymptotic behavior is exponential in nature.

There is a relationship between the hyperbolic trigonometric functions and the trigonometric functions that are based on the unit circle. If t is any real number, then the point (cos t, sin t) lies on the unit circle $x^2 + y^2 = 1$. An important trigonometric identity is $\cos^2 t + \sin^2 t = 1$. The hyperbolic trigonometric functions satisfy a similar identity: $\cosh^2 t - \sinh^2 t = 1$. If t is any real number, then the point (cos t, sin t) lies on the right branch of the hyperbola $x^2 - y^2 = 1$. This can be verified by substituting the equations $\sinh x = \frac{e^x - e^{-x}}{2}$ and $\cosh x = \frac{e^x + e^{-x}}{2}$ into the expression $\cosh^2 x - \sinh^2 x$ and simplifying. It turns out that every function in trigonometry has a corresponding hyperbolic trigonometric function, and every identity in trigonometry has a corresponding identity that involves the hyperbolic trigonometric functions. The unit circle is the basis for trigonometry, and the hyperbola $x^2 - y^2 = 1$ is the basis for hyperbolic trigonometry. The geometric interpretation of the hyperbolic functions is in itself a rich topic.

We have analyzed these functions as much as we can using our pre-calculus skills. Analyzing these functions using calculus will involve taking the derivative of these functions. Differentiating the hyperbolic functions involves differentiating a sum of exponential functions. We already know how to differentiate e^x, but we have not differentiated e^{-x}.

To differentiate $f(x) = e^{-x}$, rewrite this function as $f(x) = \frac{1}{e^x}$ and apply the quotient rule:

Apply the quotient rule $\quad f'(x) = \frac{(1)'(e^x) - (1)(e^x)'}{(e^x)^2}$

Differentiate each piece $\quad f'(x) = \frac{(0)(e^x) - (1)(e^x)}{(e^x)^2}$

Factor the denominator $\quad f'(x) = \frac{-(e^x)}{(e^x)^2}$

Cancel the common e^x factors $\quad f'(x) = \frac{-\cancel{(e^x)}}{(e^x)\cancel{(e^x)}}$

Use the rules for exponents to rewrite the derivative

$$f'(x) = -\frac{1}{e^x} = -e^{-x}$$

Now that we have established that the derivative of $f(x) = e^{-x}$ is $f'(x) = -e^{-x}$, we are ready to differentiate the hyperbolic functions:

$$\sinh x = \frac{e^x - e^{-x}}{2} = \frac{1}{2}\left(e^x - e^{-x}\right) \qquad \cosh x = \frac{e^x + e^{-x}}{2} = \frac{1}{2}\left(e^x + e^{-x}\right)$$

$$(\sinh x)' = \frac{1}{2}\left(\left(e^x\right)' - \left(e^{-x}\right)'\right) \qquad (\cosh x)' = \frac{1}{2}\left(\left(e^x\right)' + \left(e^{-x}\right)'\right)$$

$$(\sinh x)' = \frac{1}{2}\left(\left(e^x\right) - \left(-e^{-x}\right)\right) \qquad (\cosh x)' = \frac{1}{2}\left(\left(e^x\right) + \left(-e^{-x}\right)\right)$$

$$(\sinh x)' = \frac{1}{2}\left(e^x + e^{-x}\right) \qquad (\cosh x)' = \frac{1}{2}\left(e^x - e^{-x}\right)$$

$$(\sinh x)' = \cosh x \qquad (\cosh x)' = \sinh x$$

The derivative of the hyperbolic sine function is the hyperbolic cosine function, and the derivative of the hyperbolic cosine function is the hyperbolic sine function! We can now differentiate combinations of exponential, hyperbolic, and power functions together using the product and quotient rules.

Example 1

Differentiate the following functions. Do not simplify.

a. $f(x) = x \cosh x$

b. $g(x) = e^x \sinh x$

Solution: Apply the product and quotient rules, where appropriate:

a. $f(x) = x \cosh x$: $f'(x) = (x)'(\cosh x) + (x)(\cosh x)'$

$f'(x) = (1)(\cosh x) + (x)(\sinh x)$

b. $g(x) = e^x \sinh x$: $g'(x) = (e^x)'(\sinh x) + (e^x)(\sinh x)'$

$g'(x) = (e^x)(\sinh x) + (e^x)(\cosh x)$

We can also find equations of tangent lines and values of x for which the tangent line is horizontal, as we will see in the next two examples.

Example 2

Where does the graph of the function $f(x) = e^x - \sinh x$ have a horizontal tangent line?

Solution: To answer this question, take the derivative and set it equal to 0:

$$f'(x) = e^x - \cosh x = 0$$

$$e^x - \tfrac{1}{2}\left(e^x + e^{-x}\right) = 0$$

$$e^x - \tfrac{1}{2}e^x - \tfrac{1}{2}e^{-x} = 0$$

$$\tfrac{1}{2}e^x - \tfrac{1}{2}e^{-x} = 0$$

$$\tfrac{1}{2}\left(e^x - e^{-x}\right) = 0$$

$$\sinh x = 0$$

$$x = 0$$

The only value of x for which $f(x) = e^x - \sinh x$ has a horizontal tangent line is $x = 0$.

Example 3

Find the equation of the line tangent to $f(x) = \frac{\sinh x + 3}{\cosh x + 2}$ at $x = 0$.

Solution: Find the point, find the slope, and use the point-slope formula. The point is (0, 1), and the derivative of this function is:

$$f'(x) = \frac{(\sinh x + 3)'(\cosh x + 2) - (\sinh x + 3)(\cosh x + 2)'}{(\cosh x + 2)^2}$$

$$f'(x) = \frac{(\cosh x)(\cosh x + 2) - (\sinh x + 3)(\sinh x)}{(\cosh x + 2)^2}$$

$$f'(x) = \frac{(1)(1+2) - (0+3)(0)}{(1+2)^2} = \frac{1}{3}$$

Now use the point-slope formula: $y - 1 = \tfrac{1}{3}(x - 0)$

$$y = \tfrac{1}{3}x + 1$$

The graphs of $f(x)=\frac{\sinh x+3}{\cosh x+2}$ and $y=\frac{1}{3}x+1$ are shown in Figure 9.4.

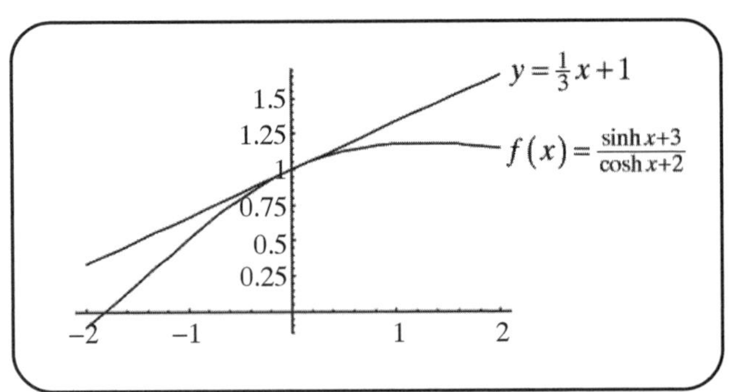

Figure 9.4.

Lesson 9-2 Review

1. Differentiate the following functions:

   a. $h(x)=\frac{\sinh x}{\cosh x}$ b. $f(x)=x^2\cosh x$

2. Find the equation of the line tangent to the following functions at the indicated point:

   a. $f(x)=\frac{x}{\cosh x}$ at $x=0$ b. $f(x)=(x^2+1)(\sinh x-2)$ at $x=0$

3. Where does the graph of the function $f(x)=x\cosh x-\sinh x$ have a horizontal tangent line?

Lesson 9-3: Trigonometric Functions

The last group of functions that we will differentiate are the trigonometric functions. Once we have a formula for the derivative of the sine and cosine functions, formulas for the rest of the trigonometric functions will follow from the quotient rule.

In Chapter 7 we derived a formula for the derivative of the sine function: $(\sin x)'=\cos x$. In the practice problems, you were given an opportunity to derive a formula for the derivative of the cosine function: $(\cos x)'=-\sin x$. Armed with this information, we can derive the formulas for the derivatives of the remaining four trigonometric functions by rewriting these functions in terms of the sine and cosine functions.

The derivatives of the trigonometric functions are summarized in the following table. To verify these formulas, write these functions in terms of the sine and cosine functions and use the quotient rule:

$(\tan x)' = \sec^2 x$	$(\cot x)' = -\csc^2 x$
$(\sec x)' = \tan x \sec x$	$(\csc x)' = -\cot x \csc x$

Example 1

Where does the function $f(x) = \sin x$ have a horizontal tangent line?

Solution: $f(x) = \sin x$ will have a horizontal tangent line wherever the derivative is equal to 0. Because $f'(x) = \cos x$ and the cosine function is zero at all odd half-integer multiples of π, $f(x) = \sin x$ will have a horizontal tangent line at $x = \frac{(2n+1)\pi}{2}$ where n is any integer.

Example 2

Where does the function $f(x) = e^x \sin x$ have a horizontal tangent line?

Solution: $f(x) = e^x \sin x$ will have a horizontal tangent line wherever the derivative is equal to 0. In order to find the derivative of this function we will need to use the product rule:

$$f'(x) = (e^x)'(\sin x) + (e^x)(\sin x)'$$

$$f'(x) = (e^x)(\sin x) + (e^x)(\cos x)$$

Now we can set the derivative equal to 0 and solve for x:

Factor e^x from both terms $\qquad (e^x)(\sin x) + (e^x)(\cos x) = 0$

Set each factor equal to 0. Because $e^x \neq 0$ for all values of x, this equation will be satisfied only where $\sin x + \cos x = 0$

$$(e^x)(\sin x + \cos x) = 0$$

Subtract $\cos x$ from both sides $\quad \sin x = -\cos x$

Divide both sides by $\cos x$ $\quad \frac{\sin x}{\cos x} = -1$

$\frac{\sin x}{\cos x} = \tan x$ $\quad \tan x = -1$

The tangent function has period π, and $\tan x = -1$ when $x = \frac{3\pi}{4}$

$$x = \frac{3\pi}{4} + n\pi$$

There are infinitely many values of x for which function $f(x) = e^x \sin x$ has a horizontal tangent line: $x = \frac{3\pi}{4} + n\pi$.

Solving problems involving horizontal tangent lines will require you to be familiar with the values of the trigonometric functions at the special angles discussed in Chapter 4.

Lesson 9-3 Review

1. Differentiate the following functions:
   a. $f(x) = \cos x \cosh x$
   b. $f(x) = \frac{\tan x}{e^x}$
2. Find the equation of the line tangent to the following functions at the indicated point:
   a. $f(x) = x \sin x$ at $x = \frac{\pi}{2}$
   b. $g(x) = \cos x \sinh x + 2$ at $x = 0$
3. Where does the graph of the function $f(x) = e^x \sec x$ have a horizontal tangent line?

Answer Key

Lesson 9-1 Review

1. a. $h'(x) = \frac{(3^x - 2^x + e^2)'(2x^4 - 7x^2 + 5) - (3^x - 2^x + e^2)(2x^4 - 7x^2 + 5)'}{(2x^4 - 7x^2 + 5)^2}$

$$h'(x) = \frac{(3^x \ln 3 - 2^x \ln 2)(2x^4 - 7x^2 + 5) - (3^x - 2^x + e^2)(8x^3 - 14x)}{(2x^4 - 7x^2 + 5)^2}$$

b. $f'(x) = \frac{(x)'(e^x+1)-(x)(e^x+1)'}{(e^x+1)^2}$

$f'(x) = \frac{(1)(e^x+1)-(x)(e^x)}{(e^x+1)^2}$

2. a. $f(x) = (x^2+1)e^x$ at $x = 0$:

The point is (0, 1), $f'(x) = (2x)e^x + (x^2+1)e^x$, so the slope is $f'(x) = 1$.
The equation of the tangent line is $y - 1 = 1(x - 0)$, or $y = x + 1$.

b. $g(x) = \frac{xe^x}{x+1}$ at $x = 0$:

The point is (0, 0), $g'(x) = \frac{[(1)e^x + xe^x](x+1) - xe^x(1)}{(x+1)^2}$, so the slope is $g'(0) = 1$.
The equation of the tangent line is $y - 0 = 1(x - 0)$, or $y = x$.

3. Solve the equation $f'(x) = (2x)e^x + x^2e^x = 0$ for x:

$e^x(2x + x^2) = 0$ at $x = 0$ or $x = -2$.

Lesson 9-2 Review

1. a. To simplify this derivative, use the identity $\cosh^2 x - \sinh^2 x = 1$:

$h(x) = \frac{(\sinh x)'(\cosh x)-(\sinh x)(\cosh x)'}{(\cosh x)^2}$

$h(x) = \frac{(\cosh x)(\cosh x)-(\sinh x)(\sinh x)}{(\cosh x)^2}$

$h(x) = \frac{(\cosh x)^2-(\sinh x)^2}{(\cosh x)^2}$

$h(x) = \frac{1}{(\cosh x)^2}$

b. $f(x) = x^2 \cosh x$: $f'(x) = (2x)\cosh x + x^2 \sinh x$

2. a. $f(x) = \frac{x}{\cosh x}$ at $x = 0$:

The point is (0, 0), $f'(x) = \frac{(1)\cosh x - x\sinh x}{\cosh^2 x}$, so the slope is $f'(0) = 1$.
The equation of the tangent line is $y - 0 = 1(x - 0)$ or $y = x$.

b. $f(x) = (x^2 + 1)(\sinh x - 2)$ at $x = 0$:

The point is (0, 2), $f'(x) = (2x)(\sinh x - 2) + (x^2 + 1)(\cosh x)$,

so the slope is $f'(x) = 1$.

The equation of the tangent line is $y + 2 = 1(x - 0)$, or $y = x - 2$.

3. Set the derivative equal to 0 and solve for x:

$f'(x) = [(1)\cosh x + x\sinh x] - \cosh x = 0$

$x\sinh x = 0$

$f(x) = x\cosh x - \sinh x$ has a horizontal tangent line at $x = 0$.

Lesson 9-3 Review

1. a. $f'(x) = (-\sin x)(\cosh x) + (\cos x)(\sinh x)$

b. $f'(x) = \frac{(\sec^2 x)(e^x) - (\tan x)(e^x)}{(e^x)^2}$

2. a. $f(x) = x\sin x$ at $x = \frac{\pi}{2}$:

The point is $\left(\frac{\pi}{2}, \frac{\pi}{2}\right)$, $f'(x) = \sin x + x\cos x$, so the slope is $f'\left(\frac{\pi}{2}\right) = 1$.

The equation of the tangent line is $y - \frac{\pi}{2} = 1\left(x - \frac{\pi}{2}\right)$, or $y = x$.

b. $g(x) = \cos x \sinh x + 2$ at $x = 0$:

The point is (0, 2), $g'(x) = (-\sin x)(\sinh x) + (\cos x)(\cosh x)$,

so the slope is $g'(0) = 1$.

The equation of the tangent line is $y - 2 = 1(x - 0)$, or $y = x + 2$.

3. Set the derivative equal to 0 and solve for x:

$f'(x) = e^x \sec x + e^x \sec x \tan x = 0$

$f'(x) = e^x \sec x(1 + \tan x) = 0$

$\tan x = -1$

$x = -\frac{\pi}{4} + n\pi$

The function has horizontal tangent lines at $x = -\frac{\pi}{4} + n\pi$.

The Chain Rule

Armed with the product and quotient rules, we can differentiate complicated functions that are combinations of power, exponential, hyperbolic, and trigonometric functions. But we cannot differentiate functions of the form $f(x) = e^{x^2}$, $g(x) = \ln x$, or $h(x) = \sin^{-1} x$. In order to differentiate these types of functions, we will need the chain rule.

Lesson 10-1: The Power Form of the Chain Rule

Technically, there is only one chain rule, but it can be thought of as taking on several forms. Looking at these different forms will help with our understanding of the chain rule in general.

The power form of the chain rule is a generalization of the product rule, and it has to do with functions of the form $f(x) = (g(x))^n$. The derivative of the function $f(x) = (g(x))^n$ is $f'(x) = n(g(x))^{n-1} g'(x)$. The coefficient n and the exponent of the function decreasing by 1 reminds us of the power rule for taking derivatives. The additional factor $g'(x)$ is the "chain" part: This derivative is tacked on to the power part via multiplication, making the chain—hence the name.

Example 1

Differentiate the following functions. Do not simplify.

a. $f(x) = (2x^2 + e^x + \sin x)^4$

b. $g(x) = \sqrt{3x + \cosh x - \tan x}$

Solution: Apply the power form of the chain rule:

a. $f(x) = (2x^2 + e^x + \sin x)^4$:

$f'(x) = 4(2x^2 + e^x + \sin x)^3 (4x + e^x + \cos x)$

b. $g(x) = \sqrt{3x + \cosh x - \tan x}$:

$g(x) = (3x + \cosh x - \tan x)^{1/2}$

$g(x) = \frac{1}{2}(3x + \cosh x - \tan x)^{-1/2}(3 + \sinh x - \sec^2 x)$

Tangent line equations are done the same way as before. The only twist with these problems is that the complexity of our functions is increasing.

Lesson 10-1 Review

1. Differentiate the following functions:

   a. $h(x) = \dfrac{\sqrt{3x}}{2 + \cos x}$

   b. $f(x) = x \sec^2 x$

2. Find the equation of the line tangent to the following functions at the indicated point:

   a. $f(x) = (3x^2 + 2x + 1)^3$ at $x = -1$

   b. $g(x) = (e^x + \tan x)^4$ at $x = 0$

3. Where does the graph of the function $f(x) = (x^2 + 5x + 10)^4$ have a horizontal tangent line?

Lesson 10-2: The General Chain Rule

We have learned the product rule, which enables us to differentiate *products* of functions. We have also discussed the quotient rule, which enables us to differentiate *ratios* of functions. The power form of the chain rule is just one specific example of the general chain rule. In general, the chain rule enables us to take the derivative of the *composition* of functions.

The derivative of the composition of two functions, $(f \circ g)(x)$ or $f(g(x))$, is given by $f'(g(x))g'(x)$. At first, you may want to write out the functions that are being composed, differentiate each function separately, and then put the pieces together. After you become more familiar with the process, you will be able to combine steps.

Example 1

Differentiate the following functions. Do not simplify.

a. $f(x) = \sin(x^2 + 2x + 1)$

b. $f(x) = e^{x^2+2x}$

Solution: Write each function as the composition of two functions, differentiate each piece, and put the pieces together.

a. $f(x) = \sin(x^2 + 2x + 1)$: Let $g(x) = \sin x$ and $h(x) = x^2 + 2x + 1$.

Then $f(x) = g(h(x))$, and $f'(x) = g'(h(x))h'(x)$.

Differentiate $g(x)$ and $h(x)$:

$g'(x) = \cos x$ and $h'(x) = 2x + 2$.

Putting everything together, we have:

$f'(x) = g'(h(x))h'(x) = \cos(x^2 + 2x + 1) \cdot (2x + 2)$.

b. $f(x) = e^{x^2+2x}$: Let $g(x) = e^x$ and $h(x) = x^2 + 2x$.

Then $f(x) = g(h(x))$, and $f'(x) = g'(h(x))h'(x)$.

Differentiate $g(x)$ and $h(x)$:

$g'(x) = e^x$ and $h'(x) = 2x + 2$.

Putting everything together, we have:

$$f'(x) = g'(h(x))h'(x) = \left(e^{x^2+2x}\right)(2x+2).$$

Mixing the product, quotient, and chain rules together can create some complicated derivatives. Applying the rules carefully is the key to working out these problems successfully. We will continue to practice finding equations of tangent lines. (If you are getting tired of finding tangent line equations, rest assured that we will discuss other applications of the derivative in the next four chapters.)

Example 2

Where does the function $f(x)=xe^{x^2-3x}$ have a horizontal tangent line?

Solution: Take the derivative, set it equal to 0, and solve for x. You will need to use the product rule and the chain rule:

$$f'(x)=(x)'\left(e^{x^2-3x}\right)+(x)\left(e^{x^2-3x}\right)'$$

$$f'(x)=(1)\left(e^{x^2-3x}\right)+(x)\left(\left(e^{x^2-3x}\right)(2x-3)\right)$$

Set the derivative equal to 0 and solve for x:

$$\left(e^{x^2-3x}\right)+(x)\left(\left(e^{x^2-3x}\right)(2x-3)\right)=0$$

Factor the e^{x^2-3x} term $\quad \left(e^{x^2-3x}\right)(1+(x)(2x-3))=0$

Distribute the x $\quad \left(e^{x^2-3x}\right)\left(1+2x^2-3x\right)=0$

Factor the quadratic expression $\left(e^{x^2-3x}\right)(2x-1)(x-1)=0$

Set each factor equal to 0 and solve for x:

$$(2x-1)=0 \text{ or } (x-1)=0$$

$$x=\tfrac{1}{2} \text{ or } x=1$$

The graph of the function $f(x)=xe^{x^2-3x}$ is shown in Figure 10.1.

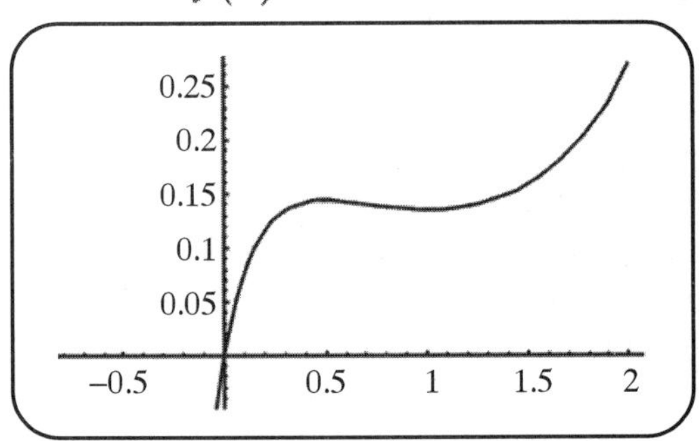

Figure 10.1.

Lesson 10-2 Review

1. Differentiate the following functions:

   a. $f(x) = \tan(x^2 + 1) \cdot \cosh(2x + e^x)$

   b. $g(x) = \dfrac{e^{x^2+x}}{(x^2+1)}$

2. Find the equation of the line tangent to the following functions at the indicated point:

   a. $f(x) = e^{x^2+2x}$ at $x = 0$

   b. $g(x) = \sinh(x^2 - 4x + 3)$ at $x = 1$

3. Where does the graph of the function $f(x) = \cosh(x^2 + 3x - 4)$ have a horizontal tangent line?

Lesson 10-3: Implicit Differentiation

When we talk about taking the derivative of a function, the derivative is taken with respect to the independent variable. If y is a function of x, then x is the independent variable and y is the dependent variable. The derivative of y with respect to x is written $\frac{dy}{dx}$ or y'. The derivative of x with respect to x can be written $\frac{dx}{dx}$, and $\frac{dx}{dx} = 1$. In all of the problems we have encountered to date, we have been able to write an explicit formula for the function. An **explicit** representation of a function is an equation in which the dependent variable is written on the left side of the equation, and the expression on the right side of the equation only involves the independent variable. Functions such as $f(x) = e^x + x$, $g(x) = \tan(x^2 + 1)$, and $y = 3x + 2$ are examples of explicit equations. Taking the derivative of these functions is now child's play.

There are times when a function is defined implicitly. **Implicit functions** are equations that involve the dependent and the independent variable throughout, and it may not be possible to write the dependent variable explicitly as a function of the independent variable. For example, the function that satisfies the equation $e^y + y = x$ cannot be written in the form $y = f(x)$. This function is certainly worth analyzing, and it has tangent lines that may be worth knowing. In order to differentiate this function, we will need to use the chain rule. Because the function is defined implicitly, the odds are good that the derivative will also have to be written implicitly, meaning that the derivative will be in terms of both x and y.

In order to differentiate an implicit equation, we need to differentiate each term in the equation. If a term only involves the independent variable, we can differentiate it using the techniques that we discussed in the past few chapters. Any terms that involve both the dependent and the independent variables must be approached using the product, quotient, or chain rules, with the added note that the derivative of y with respect to x will be written y'.

Example 1

Find y' for the following:

a. $xy + \sin y = 4$

b. $ye^x + xe^y = 1$

Solution: Differentiate each term in the equation and then solve for y':

a. $xy + \sin y = 4$: $(x \cdot y)' + (\sin y)' = (4)'$

Use the product rule to differentiate the first term

$$\left((x)' \cdot (y) + (x) \cdot (y)'\right) + (\sin y)' = (4)'$$

Differentiate each term $\left((1) \cdot (y) + (x) \cdot (y')\right) + (\cos y)y' = 0$

Keep all terms that involve y' on the left; move all other terms to the right $xy' + (\cos y)y' = -y$

Factor out y' $y'(x + \cos y) = -y$

Divide both sides of the equation by $(x + (\cos y))$

$$y' = \frac{-y}{(x+\cos y)}$$

b. $ye^x + xe^y = 1$: $(y \cdot e^x)' + (x \cdot e^y)' = (1)'$

Use the product rule to differentiate the terms on the left

$$\left((y)'\,(e^x) + (y)\,(e^x)'\right) + \left((x)'\,(e^y) + (x)\,(e^y)'\right) = (1)'$$

Differentiate each term

$$\left(y'(e^x) + (y)e^x\right) + \left((1)(e^y) + (x)(e^y y')\right) = 0$$

Keep all terms that involve y' on the left; move all other terms to the right $\quad y'e^x + xe^y y' = -ye^x - e^y$

Factor out y' $\quad y'(e^x + xe^y) = -ye^x - e^y$

Divide both sides of the equation by $(e^x + xe^y)$

$$y' = \frac{-ye^x - e^y}{\left(e^x + xe^y\right)}$$

Example 2

Find the equation of the line tangent to the graph of $e^y + xy = x$ at the point (1, 0).

Solution: We are given the point that the tangent line passes through. All we need is the slope, which means we have to take the derivative and evaluate it when $x = 1$ and $y = 0$:

$$(e^y)' + (x \cdot y)' = (x)'$$

Differentiate each term $\quad (e^y)' + \left((x)'(y) + (x)(y)'\right) = (x)'$

Use the product rule and the chain rule $\quad e^y y' + \left((1)(y) + (x)(y')\right) = (1)$

Move the terms that don't involve y' to the right side of the equation

$$e^y y' + xy' = 1 - y$$

Factor out y' $\quad y'(e^y + x) = 1 - y$

Divide both sides of the equation by $(e^y + x)$

$$y' = \frac{1-y}{\left(e^y + x\right)}$$

Evaluate the derivative when $x = 1$ and $y = 0$

$$y' = \frac{1-0}{\left(e^0 + 1\right)} = \frac{1}{2}$$

The point is (1, 0) and the slope is $\frac{1}{2}$, so we can use the point-slope formula:

$$y - 0 = \tfrac{1}{2}(x - 1)$$

$$y = \tfrac{1}{2}x - \tfrac{1}{2}$$

Lesson 10-3 Review

1. Find y':
   a. $\sin(x + y) = \cos y + x$
   b. $e^y \tan y = x$

2. Find the equation of the line tangent to the function $y \cos(x + y) = x + 2$ at the point $(-1, 1)$.

Lesson 10-4: Inverse Functions

The chain rule will enable us to differentiate inverse functions without actually having to find the inverse function itself. Remember that $y = f^{-1}(x)$ means that $x = f(y)$. If we can write an explicit equation for the inverse function, then we will probably be able to differentiate it without any problems, as we will see in our first example.

Example 1

If $f(x) = 3x + 2$, evaluate $(f^{-1})'(2)$.

Solution: Find an equation for $f^{-1}(x)$, take its derivative, and evaluate it at $x = 2$. First, find $f^{-1}(x)$ by switching x and y and then solving for y:

$$y = 3x + 2$$

Switch x and y $\quad x = 3y + 2$

Solve for y $\quad y = \frac{1}{3}(x - 2)$

Now that we have an equation for $f^{-1}(x)$, we can differentiate the function and evaluate the derivative at $x = 2$:

$$f^{-1}(x) = \tfrac{1}{3}(x - 2)$$

$$(f^{-1})'(x) = \tfrac{1}{3}$$

$$(f^{-1})'(2) = \tfrac{1}{3}$$

We can also find the equation of a line tangent to the inverse of a function, but if an explicit formula for the inverse function cannot be found, we will need another approach. The function $y = f^{-1}(x)$ means the same thing as $x = f(y)$. We can either differentiate $y = f^{-1}(x)$ (if we can actually solve for the inverse of the function), or apply the chain rule and differentiate the equation $x = f(y)$ with respect to x:

$$(x)' = (f(y))'$$

$$1 = (f'(y))y'$$

$$y' = \frac{1}{f'(y)}$$

Now, remember that this all started with the equation $y = f^{-1}(x)$, which is equivalent to $x = f(y)$. We can replace y with $f^{-1}(x)$ in our formula for y':

$$y' = \frac{1}{f'(y)} = \frac{1}{f'(f^{-1}(x))}$$

Evaluating the derivative of the inverse function at a particular point $x = a$ involves finding $f^{-1}(a)$ and then evaluating the expression $\frac{1}{f'(f^{-1}(x))}$. Let's take a specific example and see how it works.

Example 2

Evaluate $(g^{-1})'(2)$ if $g(x) = x^3 + 1$.

Solution: We will solve this problem using both methods. First, we will use the formula $y' = \frac{1}{f'(f^{-1}(x))}$. Then we will solve for the inverse function, which in this case *is* possible, and then take the derivative.

If $g(x) = x^3 + 1$, then $g^{-1}(2)$ is the value of x that g maps to 2. Finding $g^{-1}(2)$ is equivalent to solving the equation $x^3 + 1 = 2$ for x:

$$x^3 + 1 = 2$$

$$x^3 = 1$$

$$x = 1$$

So $g^{-1}(2) = 1$.

Next, evaluate $(g^{-1})'(2)$: $(g^{-1})'(2) = \frac{1}{g'(g^{-1}(2))} = \frac{1}{g'(1)}$.

We will need to find $g'(1)$:

$$g(x) = x^3 + 1$$
$$g'(x) = 3x^2$$
$$g'(1) = 3(1)^2 = 3$$

$$\left(g^{-1}\right)'(2) = \frac{1}{g'\left(g^{-1}(2)\right)} = \frac{1}{g'(1)} = \frac{1}{3}$$

Now we will solve this problem the old fashioned way. Find the inverse function and take its derivative:

$$y = x^3 + 1$$

Switch x and y $\qquad x = y^3 + 1$

Solve for y $\qquad y = \sqrt[3]{x-1}$

The inverse function is $g^{-1}(x) = \sqrt[3]{x-1}$. Now we can take the derivative of $g^{-1}(x)$ and evaluate it at $x = 2$:

$$\left(g^{-1}(x)\right)' = \left(\sqrt[3]{x-1}\right)'$$

$$\left(g^{-1}(x)\right)' = \left((x-1)^{1/3}\right)'$$

$$\left(g^{-1}(x)\right)' = \tfrac{1}{3}(x-1)^{-2/3}(1)$$

$$\left(g^{-1}\right)'(2) = \tfrac{1}{3}(2-1)^{-2/3}(1) = \tfrac{1}{3}$$

Both approaches yield the same result. Perhaps you prefer one method over the other. In situations where you can solve for the inverse function directly, you can use either method, but you *must* use the chain rule if you cannot solve for the inverse function directly. This method will be particularly useful when we differentiate the natural logarithmic function and the inverse trigonometric functions, as we will see in the next two lessons.

Lesson 10-4 Review

1. Evaluate $(f^{-1})'(1)$ if $f(x) = e^x + x$.

Lesson 10-5: Logarithmic Functions

Finding the derivative of the inverse of a function seems rather complicated. The inverse of a function takes some getting used to, but after you work a few problems you should get the hang of it. Differentiating some familiar inverses of functions may help you become more comfortable with the derivatives of inverse functions in general. The first inverse function that we will examine is the natural logarithmic function. In this lesson we will find a formula for the derivative of the function $f(x) = \ln x$.

For the function $y = \ln x$, remember that the natural logarithmic function is the inverse of the exponential function. The equation $y = \ln x$ is equivalent to the equation $e^y = x$. We can differentiate all of the terms in this equation with respect to x (using the chain rule) and then solve for y':

$$(e^y)' = (x)'$$

Differentiate each term; use the chain rule to differentiate e^y $\qquad e^y y' = 1$

Divide both sides of the equation by e^y $\qquad y' = \frac{1}{e^y}$

Use that fact that $e^y = x$ to simplify this equation $\qquad y' = \frac{1}{x}$

In other words, the derivative of the function $y = \ln x$ is just $y' = \frac{1}{x}$.

We can use the product rule, the quotient rule, and the chain rule to differentiate more complicated functions that involve the natural logarithmic function.

Example 1

Differentiate the following functions. Do not simplify.

a. $f(x) = e^{x^2} \ln x$

b. $h(x) = \ln(3x^2 + 2x + 1)$

Solution: Apply the product, quotient, and chain rules:

a. $f(x) = e^{x^2} \ln x$: $f'(x) = \left(e^{x^2}\right)'(\ln x) + \left(e^{x^2}\right)(\ln x)'$

$$f'(x) = \left(e^{x^2} 2x\right)(\ln x) + \left(e^{x^2}\right)\left(\frac{1}{x}\right)$$

b. $h(x) = \ln(3x^2 + 2x + 1)$:

To apply the chain rule, write $f(x) = \ln x$ and $g(x) = 3x^2 + 2x + 1$. Then $h(x) = f(g(x))$ and $h'(x) = f'(g(x))g'(x)$.

$$h'(x) = \frac{1}{(3x^2+2x+1)}(6x+2)$$

In general, the derivative of the function $f(x) = \ln(g(x))$ is given by the formula $f'(x) = \left(\frac{1}{g(x)}\right)g'(x)$. Sometimes it will be written $f'(x) = \frac{g'(x)}{g(x)}$. This is a very useful formula, and it is worth remembering. Along the same line, the derivative of the function $f(x) = e^{g(x)}$ is given by the formula $f'(x) = (e^{g(x)})g'(x)$. These are two special cases of the chain rule, and functions of this format are surprisingly common in science.

Example 2

Where does the function $f(x) = x \ln x$ have a horizontal tangent line?

Solution: Take the derivative, set it equal to 0 and solve for x:

$$f'(x) = (x)'(\ln x) + (x)(\ln x)'$$

$$f'(x) = (1)(\ln x) + (x)\left(\frac{1}{x}\right)$$

$$f'(x) = \ln x + 1$$

Now set the derivative equal to 0 and solve for x:

$$\ln x + 1 = 0$$

$$\ln x = -1$$

$$e^{-1} = x$$

The function $f(x) = x \ln x$ has a horizontal tangent line at $x = \frac{1}{e}$.

Example 3

Use the chain rule to derive a formula for the derivative of the function $f(x) = \log_2 x$.

Solution: Rewrite the equation in exponential form and then differentiate all terms in the equation using the chain rule:

$y = \log_2 x$

Rewrite the equation in exponential form $x = 2^y$

Differentiate both terms in the equation with respect to x. $(x)' = (2^y)'$

$1 = 2^y y' \ln 2$

Solve for y' $y' = \frac{1}{2^y \ln 2}$

Use the fact that $x = 2^y$ to simplify the equation $y' = \frac{1}{x \ln 2}$

In general, the derivative of the function $f(x) = \log_b x$ is $f'(x) = \frac{1}{x \ln b}$.

Lesson 10-5 Review

1. Differentiate the following functions:
   a. $f(x) = \ln(e^x + \cos x)$
   b. $g(x) = \frac{\ln x}{x^2+1}$

2. Find the equation of the line tangent to the function $f(x) = \ln(2x - 1)$ at $x = 1$.

3. Where does the function $f(x) = \frac{\ln x}{x}$ have a horizontal tangent line?

Lesson 10-6: Inverse Trigonometric Functions

This last application of the chain rule involves finding formulas for the derivative of the inverse trigonometric functions. We will start with the arcsine function.

To differentiate the function $y = \arcsin x$, we will first rewrite the equation as $x = \sin y$. Now we can use the chain rule to differentiate both sides of the equation with respect to x and solve for y':

$(x)' = (\sin y)'$

$1 = (\cos y)y'$

$y' = \frac{1}{\cos y}$

Technically, this formula is correct, but it is not helpful. Our goal is to find a formula for the derivative of the arcsine function in terms of x.

In order to write $y' = \frac{1}{\cos y}$ explicitly as a function of x, we need to revisit a right triangle. Figure 10.2 shows a right triangle with angle y. Recall that in the equation $y = \arcsin x$, y is the angle whose sine is x. From the right triangle shown in Figure 10.2, we see that $\sin y = x$ and, using the Pythagorean theorem, we see that $\cos y = \sqrt{1-x^2}$. Therefore:

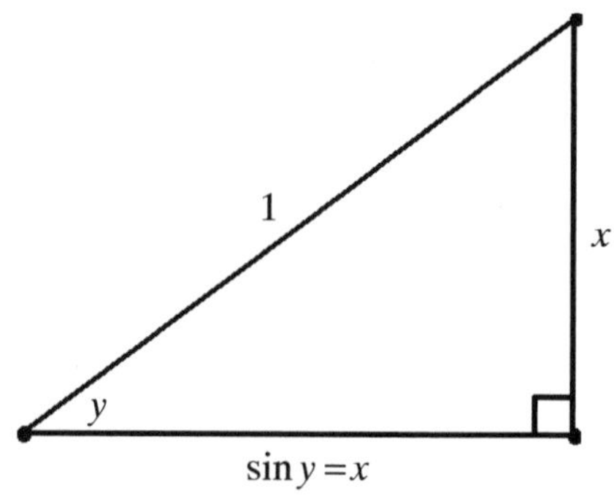

Figure 10.2.

$$y' = \frac{1}{\cos y} = \frac{1}{\sqrt{1-x^2}}$$

The derivative of the arcsine function does not involve any trigonometric functions at all!

We can apply the same technique to derive a formula for the derivative of the other inverse trigonometric functions. The three main inverse trigonometric functions and their derivatives are summarized in the following table. It is not worth memorizing these derivatives, because their derivation is simply an application of the chain rule and a little trigonometry. I would recommend working out their derivations so that you can come up with them whenever you need them.

Function	**Derivative**
$y = \arcsin x$	$y' = \frac{1}{\sqrt{1-x^2}}$
$y = \arccos x$	$y' = \frac{-1}{\sqrt{1-x^2}}$
$y = \arctan x$	$y' = \frac{1}{x^2+1}$

We can, of course, differentiate increasingly complicated functions, as we will see in the next example.

Example 1

Differentiate the following functions:

a. $f(x) = \arctan(3x^2 + 2x + 1)$

b. $g(x) = e^{\arcsin x}$

Solution: Apply the product rule and the chain rule:

a. $f(x) = \arctan(3x^2 + 2x + 1)$: $f'(x) = \left(\frac{1}{(3x^2+2x+1)^2+1}\right)(6x+2)$

b. $g(x) = e^{\arcsin x}$: $g'(x) = \left(e^{\arcsin x}\right)\frac{1}{\sqrt{1-x^2}}$

Lesson 10-6 Review

Differentiate the following functions:

1. $f(x) = x \arccos(2x + 1)$
2. $g(x) = \ln\left((\arctan x)^2 + e^x\right)$

Lesson 10-7: Parametric Derivatives

A parametric representation of a graph is a system of equations $\begin{cases} x = x(t) \\ y = y(t) \end{cases}$ where t is the independent variable and both x and y are the dependent variables. As t changes, both x and y change, and the graph is traced out. Parametric equations can be used to represent *relations* using *functions*. For example, the circle $x^2 + y^2 = 1$ is not a function. The unit circle fails the vertical line test for functions because it does not satisfy the requirement that for every value of x there corresponds a *unique* value of y. The graph of the unit circle does, however, have tangent lines that can be useful when studying rotational motion. Writing the unit circle using parametric equations will enable us to compute the slope of lines tangent to the unit circle.

One of the things to realize about parametric equations is that there is a direction in which the graph is traced out. Consider the parametric representation of the unit circle $\begin{cases} x = \cos t \\ y = \sin t \end{cases}$. This system of equations represents the unit circle because of the identity $\cos^2 t + \sin^2 t = 1$. The point corresponding to $t = 0$ is the point $(\cos 0, \sin 0)$, or $(1, 0)$. As t increases from 0 to $\frac{\pi}{2}$, x decreases from 1 to 0 and y increases from 0 to 1, and the unit circle is traced in a counter-clockwise direction. The entire circle is traced out as t ranges from 0 to 2π, or one period of the sine or cosine function.

Parametric representations for a graph are not unique. The system of equations $\begin{cases} x = \sin t \\ y = \cos t \end{cases}$ also traces out the unit circle. The point corresponding to $t = 0$ is the point $(\sin 0, \cos 0)$, or $(0, 1)$, and the unit circle is traced out clockwise. The system of equations $\begin{cases} x = \cos 2t \\ y = \sin 2t \end{cases}$ also traces out the unit circle, but it traces it out "faster" in the sense that the entire circle is completed as t ranges from 0 to π. When we calculate the slope of the line tangent to a graph described parametrically, the slope must be independent of the particular parametric representation used.

The notation for a derivative that reminds us of fractions, $\frac{dy}{dx}$, can motivate the process for taking the derivative of a function (or relation) that is described using a parametric representation. We can write $\frac{dy}{dx} = \frac{\frac{dy}{dt}}{\frac{dx}{dt}}$ and realize that the derivative of a relation defined using a parametric representation is the ratio of the derivatives of the equations in the parametric representation.

Example 1

Find the equation of the line tangent to the parametric curve defined by $\begin{cases} x = \cos 2t \\ y = \sin 3t \end{cases}$ at $t = \frac{\pi}{12}$.

Solution: We need a point and a slope.

The point is $\left(\cos 2\left(\frac{\pi}{12}\right), \sin 3\left(\frac{\pi}{12}\right)\right)$, which evaluates to $\left(\cos\left(\frac{\pi}{6}\right), \sin\left(\frac{\pi}{4}\right)\right)$ or $\left(\frac{\sqrt{3}}{2}, \frac{\sqrt{2}}{2}\right)$. The slope of the tangent line is given by the formula $\frac{dy}{dx} = \frac{\frac{dy}{dt}}{\frac{dx}{dt}}$.

Differentiating both parametric equations, we have

$$\frac{dx}{dt} = -2\sin 2t \text{ and } \frac{dy}{dt} = 3\cos 3t.$$

So we have $\frac{dy}{dx} = \frac{\frac{dy}{dt}}{\frac{dx}{dt}} = \frac{3\cos 3t}{-2\sin 2t}$.

We can evaluate this derivative when $t = \frac{\pi}{12}$:

$$\left.\frac{dy}{dx}\right|_{t=\frac{\pi}{12}} = \frac{3\cos 3\left(\frac{\pi}{12}\right)}{-2\sin 2\left(\frac{\pi}{12}\right)} = \frac{3\cos\left(\frac{\pi}{4}\right)}{-2\sin\left(\frac{\pi}{6}\right)} = \frac{3\left(\frac{\sqrt{2}}{2}\right)}{-2\left(\frac{1}{2}\right)} = -\frac{3\sqrt{2}}{2}.$$

Using the point-slope formula, we have:

$$y - \frac{\sqrt{2}}{2} = -\frac{\sqrt{3}}{2}\left(x - \frac{\sqrt{3}}{2}\right)$$

$$y = -\frac{\sqrt{3}}{2}x + \left(\frac{3}{4} + \frac{\sqrt{2}}{2}\right)$$

Lesson 10-7 Review

1. Find the equation of the line tangent to the parametric curve defined by $\begin{cases} x = \sin 2t \\ y = \cos t \end{cases}$ at $t = \frac{\pi}{6}$.

Answer Key

Lesson 10-1 Review

1. a. $h'(x) = \frac{\left((3x)^{\frac{1}{2}}\right)'(2+\cos x) - (3x)^{\frac{1}{2}}(2+\cos x)'}{(2+\cos x)^2}$

$h'(x) = \frac{\left(\frac{1}{2}(3x)^{-\frac{1}{2}}(3)\right)(2+\cos x) - (3x)^{\frac{1}{2}}(-\sin x)}{(2+\cos x)^2}$

b. $f'(x) = (x)'(\sec^2 x) + (x)(\sec^2 x)'$

$f'(x) = (1)(\sec^2 x) + (x)[(2\sec x)(\sec x \tan x)]$

2. a. $f(x) = (3x^2 + 2x + 1)^3$ at $x = -1$:
   The point is $(-1, 8)$, $f'(x) = 3(3x^2 + 2x + 1)^2(6x + 2)$, so the slope is $f'(-1) = -48$.
   The equation of the tangent line is $y - 8 = -48(x + 1)$, or $y = -48x - 40$.
   b. $g(x) = (e^x + \tan x)^4$ at $x = 0$:
   The point is $(0, 1)$, $g'(x) = 4(e^x + \tan x)^3 (e^x + \sec^2 x)$, so the slope is $g'(0) = 8$.
   The equation of the tangent line is $y - 1 = 8(x - 0)$, or $y = 8x + 1$.
3. Set the derivative equal to 0 and solve for x:
   $f'(x) = 4(x^2 + 5x + 10)^3(2x + 5) = 0$ at $x = -\frac{5}{2}$.

Lesson 10-2 Review

1. a. $f'(x) = (\tan(x^2 + 1))'(\cosh(2x + e^x)) + (\tan(x^2 + 1))(\cosh(2x + e^x))'$
   $f'(x) = (\sec^2(x^2 + 1)\cdot(2x))(\cosh(2x + e^x)) + (\tan(x^2 + 1))(\sinh(2x + e^x)\cdot(2 + e^x))$

   b. $g'(x) = \dfrac{\left(e^{x^2+x}\right)'\left(x^2+1\right) - \left(e^{x^2+x}\right)\left(x^2+1\right)'}{\left(x^2+1\right)^2}$

   $g'(x) = \dfrac{\left(\left(e^{x^2+x}\right)(2x+1)\right)\left(x^2+1\right) - \left(e^{x^2+x}\right)(2x)}{\left(x^2+1\right)^2}$

2. a. $f(x) = e^{x^2+2x}$ at $x = 0$:

   The point is $(0, 1)$, $f'(x) = \left(e^{x^2+2x}\right)(2x + 2)$, so the slope is $f'(0) = 2$.

   The equation of the tangent line is $y - 1 = 2(x - 0)$, or $y = 2x + 1$.
   b. $g(x) = \sinh(x^2 - 4x + 3)$ at $x = 1$:
   The point is $(1, 0)$, $g'(x) = (\cosh(x^2 - 4x + 3))(2x - 4)$, so the slope is $g'(0) = -4$.
   The equation of the tangent line is $y - 0 = -4(x - 1)$, or $y = -4x + 4$.
3. Set the derivative equal to 0 and solve for x: $f'(x) = (\sinh(x^2 + 3x - 4))(2x + 3) = 0$. Either $\sinh(x^2 + 3x - 4) = 0$ or $(2x + 3) = 0$. The hyperbolic sine function is 0 only when its argument is 0, which means that $(x^2 + 3x - 4) = 0$. The horizontal asymptotes occur at $x = -4$, $x = 1$, or $x = -\frac{3}{2}$.

Lesson 10-3 Review

1. a. Find y' if $\sin(x + y) = \cos y + x$:
   $(\sin(x + y))' = (\cos y)' + (x)'$

Differentiate each term	$\cos(x+y)(1+y') = (-\sin y)y' + 1$
Distribute $\cos(x+y)$	$\cos(x+y) + y'\cos(x+y) = (-\sin y)y' + 1$

Keep all terms that involve y' on the left; move all other terms to the right

$$(\sin y)y' + y'\cos(x+y) = 1 - \cos(x+y)$$

Factor out y' $\quad y'(\sin y + \cos(x+y)) = 1 - \cos(x+y)$

Divide both sides of the equation by $(\sin y + \cos(x+y))$ $\quad y' = \frac{1-\cos(x+y)}{(\sin y+\cos(x+y))}$

b. $e^y \tan y = x$:

$(e^y \tan y)' = (x)'$

Differentiate each term $\quad (e^y y')(\tan y) + (e^y)(\sec^2 y y') = 1$

Factor y' from both terms on the left $\quad y'(e^y)(\tan y + \sec^2 y) = 1$

Divide by $(e^y)(\tan y + \sec^2 y)$ $\quad y' = \frac{1}{(e^y)(\tan y+\sec^2 y)}$

2. The point is given: (-1, 1). Differentiate the function and substitute $x = -1$ and $y = 1$ into the differentiated equation and solve for y':

$(y\cos(x+y))' = (x+2)'$

$y'\cos(x+y) - y\sin(x+y)(1+y') = 1$

$y'\cos(-1+1) - 1\cdot\sin(-1+1)(1+y') = 1$

$y'\cdot 1 - 1\cdot 0\cdot(1+y') = 1$

$y' = 1$

The slope is 1, so the equation of the tangent line is $y - 1 = 1(x+1)$, or $y = x + 2$.

Lesson 10-4 Review

1. $f(x) = e^x + x = 1$ at $x = 0$, so $f^{-1}(1) = 0$

$f'(x) = e^x + 1$, so we have:

$$\left(f^{-1}\right)'(1) = \frac{1}{f'\left(f^{-1}(1)\right)} = \frac{1}{f'(0)} = \frac{1}{e^0+1} = \frac{1}{2}$$

Lesson 10-5 Review

1. a. $f'(x) = \left(\frac{1}{(e^x+\cos x)}\right)\left(e^x - \sin x\right)$

b. $g'(x) = \frac{(\ln x)'\left(x^2+1\right)-(\ln x)\left(x^2+1\right)'}{\left(x^2+1\right)^2}$

$g'(x) = \frac{\left(\frac{1}{x}\right)\left(x^2+1\right)-(\ln x)(2x)}{\left(x^2+1\right)^2}$

2. The point is $(1, 0)$, $f'(x) = \left(\frac{1}{(2x-1)}\right)(2)$, so the slope is $f'(1) = 2$.

   The equation of the tangent line is $y - 0 = 2(x - 1)$, or $y = 2x - 2$.

3. Set the derivative equal to 0 and solve for x:

$$f'(x) = \frac{(\ln x)'(x) - (\ln x)(x)'}{x^2}$$

$$f'(x) = \frac{\left(\frac{1}{x}\right)(x) - (\ln x)(1)}{x^2} = 0$$

$$f'(x) = \frac{1 - \ln x}{x^2} = 0$$

$1 - \ln x = 0$

$\ln x = 1$

$x = e$

The function has a horizontal tangent line at $x = e$.

Lesson 10-6 Review

1. $f'(x) = (x)'(\arccos(2x+1)) + (x)(\arccos(2x+1))'$

$$f'(x) = (1)(\arccos(2x+1)) + (x)\left(\frac{-1}{\sqrt{1-(2x+1)^2}}(2)\right)$$

2. $g'(x) = \left(\frac{1}{\left((\arctan x)^2 + e^x\right)}\right)\left(2(\arctan x)\left(\frac{1}{1+x^2}\right) + e^x\right)$

Lesson 10-7 Review

1. The point is $\left(\sin 2\left(\frac{\pi}{6}\right), \cos\left(\frac{\pi}{6}\right)\right)$, or $\left(\frac{\sqrt{3}}{2}, \frac{\sqrt{3}}{2}\right)$.

$$\frac{dx}{dt} = 2\cos 2t,\ \left.\frac{dx}{dt}\right|_{t=\frac{\pi}{6}} = 2\cos 2\left(\frac{\pi}{6}\right) = 2\cos\left(\frac{\pi}{3}\right) = 1$$

$$\frac{dy}{dt} = -\sin t,\ \left.\frac{dx}{dt}\right|_{t=\frac{\pi}{6}} = -\sin\left(\frac{\pi}{6}\right) = \frac{1}{2}$$

$$\frac{dy}{dx} = \frac{\frac{dy}{dt}}{\frac{dx}{dt}} = \frac{\frac{1}{2}}{1} = \frac{1}{2}$$

The slope is $\frac{1}{2}$, so the equation of the tangent line is $y - \frac{\sqrt{3}}{2} = \frac{1}{2}\left(x - \frac{\sqrt{3}}{2}\right)$, or

$y = \frac{1}{2}x + \frac{\sqrt{3}}{4}$.

Applications of the Derivative

We have discussed the derivative of a variety of functions, including power functions, exponential functions, logarithmic functions, hyperbolic functions, and trigonometric functions. We have also discussed the product rule, the quotient rule, and the chain rule. We can now take the derivative of most of the functions that you will encounter in physics, chemistry, engineering, and business. If you do encounter a function that we have not discussed, you can always use the definition of the derivative and evaluate the limit algebraically or numerically.

Now that you have mastered the techniques involved in taking derivatives, it's time to discuss the many applications of the derivative. We have already found equations of tangent lines, and we have found points in the domain where the tangent line is horizontal. We begin this chapter with a discussion of why tangent lines are so important. From there we will apply the concept of the derivative to motion and rates of change.

Lesson 11-1: Using the Tangent Line

Many of my students believe that the applications of calculus are archaic. I admit that calculators and computers can solve many calculus problems much more quickly and accurately than we can by hand. I also realize that calculators and computers do not have an inherent intelligence. They are only as capable as the person who writes their programs, and there are many situations in which computers and calculators consistently give the *wrong* answer because of flaws in their algorithms.

I advocate understanding the basic principles behind problem-solving strategies and relying less on technology for "the answer." Understanding these basic principles can be accomplished by solving somewhat simplistic problems, or problems that can easily be solved using a calculator or a computer. The goal is not to compete with our calculators or computers, but rather to understand how our technology works and become aware of its limitations.

Historically, calculus was developed when calculators were not as capable as they are now. Calculations could be performed more quickly and accurately using calculus. For example, it is not difficult to evaluate the function $f(x)=\sqrt{x}$ for values of x that are perfect squares: $f(4)=\sqrt{4}=2$ and $f(9)=\sqrt{9}=3$. We can use calculus to evaluate $f(4.2)=\sqrt{4.2}$ somewhat accurately without a calculator.

One application of calculus involves replacing a complicated function with its tangent line. A tangent line glances off the graph of a function, and in a small region centered at the point of intersection the tangent line and the graph of the function are very close to each other—so close that the function can be replaced by its tangent line in that small region. The size of the region is determined by how close you want the tangent line and the function to be. Replacing a function with its tangent line will introduce an error, and any limitations on the allowable error will determine the size of the region.

The line $y=\frac{1}{4}x+1$ is tangent to the function $f(x)=\sqrt{x}$ at $x = 4$. The graph of the function $f(x)=\sqrt{x}$ along with its tangent line at $x = 4$ is shown in Figure 11.1. When $x = 4.2$, the graphs of $f(x)=\sqrt{x}$ and $y=\frac{1}{4}x+1$ are very close to each other, and we can approximate $\sqrt{4.2}$ by evaluating

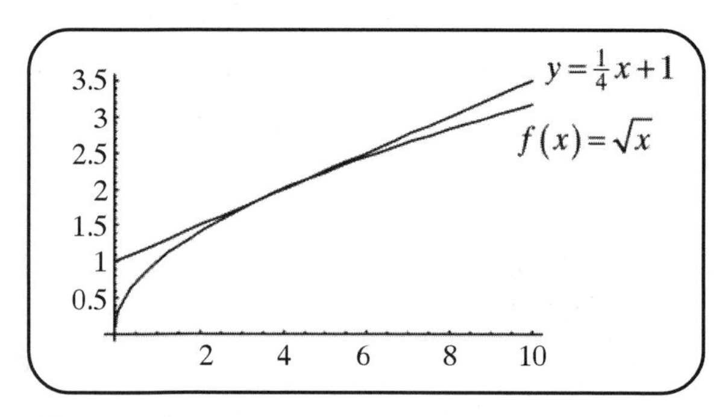

Figure 11.1.

$y = \frac{1}{4}x + 1$ at $x = 4.2$: When $x = 4.2$, $y = 2.05$. We can use this value of y to approximate $\sqrt{4.2}$: $\sqrt{4.2} \approx 2.05$. To six decimal places, $\sqrt{4.2} = 2.049390$, so our approximation is accurate to two decimal places.

The application of calculus to calculations has decreased in importance over the years because of technological advances, but there are times when using calculus to approximate functions is useful. To evaluate the function $f(x) = e^x \sin x + e^{x^2} \cosh x$ at $x = 0.01$ may take some time to enter into a calculator, but the equation of the line tangent to $f(x)$ at $x = 0$ is $y = x + 1$, and when $x = 0.01$, $y = 1.01$, so $f(0.01) \approx 1.01$. In actuality, to six decimal places, $f(0.01) = 1.010250$, and in this situation, the tangent line is easier to work with than a calculator.

A function can be approximated by its tangent line when the value of the function and its derivative are known at a nearby point. For example, we can use the line $y = \frac{1}{4}x + 1$ to approximate the function $f(x) = \sqrt{x}$ near $x = 4$. If we try to use the line $y = \frac{1}{4}x + 1$ to estimate $\sqrt{10}$, we would get $\sqrt{10} \approx \frac{1}{4}(10) + 1 = 3.5$. We would be better off approximating the function $f(x) = \sqrt{x}$ with either the tangent line at $x = 9$ or the tangent line at $x = 16$ (the two perfect squares that sandwich 10). The line tangent to the graph of $f(x) = \sqrt{x}$ at $x = 9$ is $y = \frac{1}{6}x + \frac{3}{2}$. When $x = 10$, $y = \frac{19}{6} = 3.167$, and $\sqrt{10} \approx 3.167$. The line tangent to the graph of $f(x) = \sqrt{x}$ at $x = 16$ is $y = \frac{1}{8}x + 2$. When $x = 10$, $y = 3.25$, and $\sqrt{10} \approx 3.25$.

In actuality, $\sqrt{10} = 3.162278$. The tangent line at $x = 9$ gives a better estimate of $\sqrt{10}$ than does the tangent line at $x = 16$. That is because 10 is closer to 9 than it is to 16, and the closer tangent line usually gives a better approximation. I wrote *usually*, instead of *always*, because strange things can happen.

Example 1

Suppose $f(2) = 3$ and $f'(2) = -2$. Estimate $f(2.3)$.

Solution: In order to estimate $f(2.3)$, we will need to find a tangent line equation. The only point that we know anything about is at $x = 2$, which is close to our desired target value of 2.3. The tangent line will pass through the point (2, 3), and its slope will be –2. Using the point-slope formula, we have:

$$y-3=-2(x-2)$$
$$y=-2x+7$$

Now we can use the tangent line to approximate $f(2.3)$: When $x=2.3, y=2.4$, so $f(2.3)\approx 2.4$.

Example 2

Suppose $f(4)=-3, f'(4)=-1$, and $G(x)=\frac{\ln(f(x)+4)}{x}$.

Find the line tangent to $G(x)$ at $x=4$.

Solution: The line tangent to $G(x)$ at $x=4$ will pass through the point (4, G(4)) and will have slope $G'(x)$. First, determine the point:

$$G(4)=\frac{\ln(f(4)+4)}{4}$$

$$G(4)=\frac{\ln(-3+4)}{4}=0$$

The point is (4, 0). Next, determine the slope by differentiating $G(x)$:

$$G(x)=\frac{\ln(f(x)+4)}{x}$$

$$G'(x)=\frac{\left(\ln(f(x)+4)\right)'(x)-\left(\ln(f(x)+4)\right)(x)'}{(x)^2}$$

$$G'(x)=\frac{\left(\frac{1}{f(x)+4}\right)\left(f'(x)\right)(x)-\left(\ln(f(x)+4)\right)(1)}{(x)^2}$$

Now evaluate the derivative at $x=4$:

$$G'(4)=\frac{\left(\frac{1}{f(4)+4}\right)\left(f'(4)\right)(4)-\left(\ln(f(4)+4)\right)(1)}{(4)^2}$$

$$G'(4)=\frac{\left(\frac{1}{f(4)+4}\right)\left(f'(4)\right)(4)-\left(\ln\left(f(4)+4\right)\right)(1)}{(4)^2}$$

$$G'(4)=\frac{\left(\frac{1}{-3+4}\right)(-1)(4)-\left(\ln(-3+4)\right)(1)}{(4)^2}$$

$$G'(4)=\frac{(1)(-1)(4)-\left(\ln(1)\right)(1)}{(4)^2}$$

$$G'(4)=\frac{-4-(0)(1)}{16}=-\frac{1}{4}$$

The slope is $-\frac{1}{4}$. Now use the point-slope formula to find the equation of the tangent line:

$$y-0=-\tfrac{1}{4}(x-4)$$
$$y=-\tfrac{1}{4}x+1$$

Example 3

Suppose $f(2) = 3$, $f'(2) = -2$, and $G(x) = e^{x-2} f(x) + 2$. Estimate $G(2.3)$.

Solution: In order to estimate $G(2.3)$, we will need to find the equation of a tangent line near $x = 2.3$. We have information about $f(x)$, and hence $G(x)$ at $x = 2$, which is near $x = 2.3$. The line tangent to $G(x)$ at $x = 2$ will pass through the point $(2, G(2))$ and will have slope $G'(2)$. First, find the point:

$$G(x) = e^{x-2} f(x) + 2$$
$$G(2) = e^{2-2} f(2) + 2$$
$$G(2) = e^{0} (3) + 2 = 5$$

The point is $(2, 5)$. Next, find the derivative of $G(x)$:

$$G(x) = e^{x-2} f(x) + 2$$
$$G'(x) = (e^{x-2})'(f(x)) + (e^{x-2})(f(x))'$$
$$G'(x) = (e^{x-2}(1))(f(x)) + (e^{x-2})(f'(x))$$

Now evaluate the derivative at $x = 2$:

$$G'(2) = (e^{2-2}(1))(f(2)) + (e^{2-2})(f'(2))$$

$$G'(2) = (e^{0}(1))(3) + (e^{0})(-2) = 1$$

The slope is 1. Use the point-slope formula to find the equation of the tangent line:

$$y - 5 = 1(x - 2)$$

$$y = x + 3$$

Finally, evaluate y when $x = 2.3$: When $x = 2.3$, $y = 5.3$, so $G(2.3) \approx 5.3$.

The process of approximating a function with its tangent line is referred to as finding a **local linearization** of a function. It is a linearization because we are replacing the function with a line. It is local because the most accurate results are obtained when we are close to the point where the tangent line intersects the graph of the function.

Lesson 11-1 Review

1. Find the equation of the line tangent to the graph of $f(x) = e^x \tan^{-1} x$ at $x = 0$ and use it to approximate $f(0.1)$. Compare the approximation to the actual value.
2. Suppose $f(3) = -2$ and $f'(3) = 4$. Estimate $f(2.85)$.
3. Suppose $f(4) = 5$, $f'(4) = -3$, and $G(x) = \frac{e^{x-4}}{f(x)}$. Find the line tangent to $G(x)$ at $x = 4$.
4. Suppose $f(2) = 3$, $f'(2) = -2$, and $G(x) = x \ln(f(x) - 2)$. Estimate $G(2.2)$.
5. Find the local linearization of the function $f(x) = \frac{1}{(1+2x)^4}$ at $x = 0$.

Lesson 11-2: Derivatives and Motion

A common application of the derivative has to do with motion. The derivative of a function is the rate of change of a function, and the magnitude of the rate of change is often thought of as *speed*.

When applying calculus to motion, we start with a position function. A **position function** is a function that describes the position of an object as a function of time. Because time is the independent variable, it is common to use t to represent the independent variable. If a ball is thrown up in the air, the position of the ball will depend on the time elapsed since the ball was thrown. Of course, the initial velocity that the ball was thrown with, the force (gravity) that is acting to pull the ball down to the Earth, and the initial position of the ball must also be incorporated into the equation. We can make the problem more difficult by including wind resistance and other factors, but given these initial conditions, calculus can help determine the position of the ball during its flight.

If we start with a position function, $h(t)$, of a particular object, we can obtain a velocity function $v(t)$. The velocity function $v(t)$, describes the change in distance per unit time. The change in distance per unit time is a difference quotient, and the **average velocity** over a time interval $[a, b]$ is the ratio of the change in distance to the change in time:

$$\text{average velocity} = \frac{\Delta h}{\Delta t} = \frac{h(b)-h(a)}{b-a}$$

The average velocity is the slope of the secant line between the points $(a, h(a))$ and $(b, h(b))$. The **instantaneous velocity** is the limit of this difference quotient as b approaches a. In other words:

$$\text{instantaneous velocity} = \lim_{\Delta t \to 0} \frac{\Delta h}{\Delta t} = \lim_{b \to a} \frac{h(b)-h(a)}{b-a} = h'(a)$$

The instantaneous velocity function, or the velocity function, is the *derivative* of the position function. Mathematically, it is possible for the velocity of an object to be negative. This happens when the final height of the object is lower than the initial height. Although we tend to use the words *velocity* and *speed* interchangeably, there is a distinction that scientists make between the two words. **Speed** is the magnitude of the velocity. Speed must be either positive or 0; speed cannot be negative.

Example 1

A grapefruit is thrown up in the air from an initial height of 6 feet. If the vertical height of the grapefruit relative to the ground, $h(t)$, after t seconds is given by the formula $h(t) = -16t^2 + 50t + 6$, find the height and the velocity of the grapefruit 3 seconds after it is thrown. Based on this information, is the grapefruit on its way up or down?

Solution: First, we need to get oriented. Because $h(t)$ is a measure of the vertical height of the grapefruit relative to the ground, when the grapefruit is on its way up, the change in height will be positive and its velocity will be positive. When the grapefruit is on its way down, its change in height will be negative, so its velocity will be negative. The units for velocity are the units of height divided by the units of time. Because our height is measured in feet and time is measured in seconds, the units for velocity will be feet per second. The height 3 seconds after the grapefruit is thrown is found by evaluating $h(3)$:

$$h(3) = -16(3)^2 + 50(3) + 6 = 12$$

The height 3 seconds after it is thrown is 12 feet. The velocity function is the derivative of the position function: $v(t) = h'(t) = -32t + 50$. The velocity 3 seconds after it is thrown is found by evaluating $v(3) = h'(3)$:

$$v(3) = -32(3) + 50 = -46$$

The velocity of the grapefruit is –46 feet per second. The height of the grapefruit is positive, so it is still in the air. The velocity is negative, which means that the grapefruit is on its way down.

In general, if an object is being thrown in the air, a simple model from physics will enable us to write a position function. This type of motion is often called **projectile motion.** If the initial velocity is v_0, and the object has an initial height h_0, and the gravitational force acting on the object is a constant g, then the position of the object as a function of time is given by this formula:

$$h(t) = \left(-\tfrac{1}{2}g\right)t^2 + v_0 t + h_0$$

Compare the formula given in Example 1 to this general formula. In Example 1, our units for distance were feet, and the leading coefficient was $-16 = -\frac{1}{2}(32)$. In effect, we used the gravitational constant $32\frac{\text{ft}}{\text{sec}^2}$.

The initial velocity was $50\frac{\text{ft}}{\text{sec}}$, and the initial height was 6 ft. The constant g is called the **acceleration due to gravity.** It is different on the moon than it is on the Earth. It also depends on the units that are used to measure the height. If the height is measured in feet, $g = 32\frac{\text{ft}}{\text{sec}^2}$, and if the height is measured in meters, $g = 9.8\frac{\text{m}}{\text{sec}^2}$. On the moon, $g = 5.31\frac{\text{ft}}{\text{sec}^2}$

or $g = 1.62\frac{\text{m}}{\text{sec}^2}$, depending on the units for height. We will assume that the acceleration due to gravity is the same regardless of where you are on the Earth.

Example 2

A penny is dropped from the roof of the Empire State Building, at an initial height of 1,250 feet. If it is not given any initial velocity, find the velocity of the penny when it hits the ground.

Solution: When the penny hits the ground, its height will be 0. We will need to find the time that it takes the penny to hit the ground and then evaluate the velocity function at this time. We need to create the position function. Because our height was given in feet, $g = 32$. The initial velocity is 0 and the initial height is 1,250 feet, so the position function is:

$$h(t) = \left(-\tfrac{1}{2}g\right)t^2 + v_0 t + h_0$$

$$h(t) = -16t^2 + 1250$$

First, find when the penny hits the ground. Set $h(t) = 0$ and solve for t:

$$-16t^2 + 1250 = 0$$

$$16t^2 = 1250$$

$$t^2 = \frac{1250}{16}$$

$$t = \pm 12.5\sqrt{2}$$

The negative value of time does not make sense. It could be interpreted as going back in time, which does not apply in this situation. So the penny will hit the ground $12.5\sqrt{2}$ seconds after it is dropped. Next, find the velocity function by taking the derivative of the position function:

$$h(t) = -16t^2 + 1250$$

$$v(t) = h'(t) = -32t$$

$$v\left(12.5\sqrt{2}\right) = (-32)\left(12.5\sqrt{2}\right) \approx -565.69$$

The velocity of the penny when it hits the ground is $-565.69\frac{\text{ft}}{\text{sec}}$.

For projectile motion, the model used to determine the height as a function of time is an equation of the form:

$$h(t)=\left(-\tfrac{1}{2}g\right)t^2+v_0t+h_0$$

The height function is a parabola that opens down. Its maximum value occurs at the vertex of the parabola. Finding the maximum height for projectile motion simply involves writing a quadratic model for the motion and finding the vertex. For more complicated motion, we will be able to use calculus to determine the maximum (or minimum) height.

Velocity is the derivative of the position function, and **acceleration** is the derivative of the velocity function. In other words, acceleration is the *derivative* of the *derivative* of the position function. The derivative of the derivative is called the **second derivative** and is denoted with two primes. Acceleration is the second derivative of the position function and can be written $a(t) = v'(t) = h''(t)$.

Lesson 11-3: Differentials

In Lesson 11-1 we saw that we can approximate a function using its tangent line. There is a certain amount of error involved in this approximation, and differentials can be used to determine this error.

If $y = f(x)$ is a differentiable function, then the **differential** dy is defined by $dy = f'(x)dx$. In the equation for the differential dy, both x and dx are variables, where x can be interpreted as the independent variable being measured and dx is the error in measurement. In this case, dy represents an approximation to the error in calculating y.

Example 1

The radius of a circular disk is 24 cm with a possible error in measurement of 0.2 cm. Find the area of the disk and use differentials to estimate the possible error in calculating this area.

Solution: The area of a circle is $A(r) = \pi r^2$, so:

$A(24) = \pi(24)^2 = 1{,}809.56$

The area of the disk is 1,809.56 cm^2. The maximum error can be estimated using differentials: $dA = 2\pi r dr$. The radius is 24 cm and the error, or dr, is 0.2 cm. Using this information, we have $dA = 2\pi(24)(0.2) = 30.16$. The possible error in calculating this area is 30.16 cm^2.

The **relative error** is found by evaluating the ratio $\frac{dy}{y}$. In Example 1, the possible error was 30.16 cm^2 and the calculated area was 1,809.56 cm^2. The relative error is the ratio $\frac{30.16}{1809.56} = 0.0167$, or 1.67%.

Differentials can also be used to approximate the change in a function: $\Delta y \approx dy$. Because $\Delta y = y_{final} - y_{initial}$, we have $y_{final} - y_{initial} \approx dy$ or $y_{final} \approx y_{initial} + dy$. In order to use differentials to approximate a function, first determine the values for $x_{initial}$ and dx, keeping in mind that the *target* value of x will be $x_{initial} + dx$. Evaluate $y_{initial} = f(x_{initial})$ and find dy using the equation $dy = f'(x_{initial})dx$. Put all the pieces together to estimate y_{final} using the equation $y_{final} \approx y_{initial} + dy$.

Example 2

Use differentials to approximate $\sqrt[3]{9}$.

Solution: Let $y = \sqrt[3]{x}$. Then $dy = \frac{1}{3}x^{-2/3}dx$.

If $x_{initial} = 8$ and $dx = 1$, then $y_{initial} = \sqrt[3]{8} = 2$, and $dy = \frac{1}{3}(8)^{-2/3}(1) = \frac{1}{12}$.

Using the equation $y_{final} \approx y_{initial} + dy$, we have $y_{final} \approx 2 + \frac{1}{12} = 2.083$.

Our approximation for $\sqrt[3]{9}$ is 2.083, and the actual value is 2.08008.

You have now seen how to approximate functions using tangent lines and using differentials. Both methods are equivalent, despite their apparent differences. The tangent line approximation of a function is given by this equation:

$$y = f(a) + f'(a)(x - a)$$

If $f(a) = y_{initial}$ and $(x - a) = dx$, then $y = y_{initial} + f'(a)dx$.

Because $dy = f'(a)dx$, we have $y = y_{initial} + dy$.

Starting with the tangent line equation, we can derive the equivalent equation using differentials.

Lesson 11-3 Review

1. The edge of a cube was found to be 50 cm with a possible error in measurement of 0.1 cm. Use differentials to estimate the relative error in computing the volume of the cube and the surface area of the cube.

2. Use differentials to approximate $\sqrt[5]{34}$.

Lesson 11-4: Rolle's Theorem and the Mean Value Theorem

It is common in mathematics to prove that something *can* be done, or that something *does* exist, without giving specific details on *how* to actually perform the task. These types of results are called **existence theorems.** Rolle's Theorem and the Mean Value Theorem are two examples of such mathematical results.

Rolle's Theorem

If a function $f(x)$ is continuous on a closed interval $[a, b]$ and is differentiable on the open interval (a, b), and satisfies the condition that $f(a) = f(b)$, then there exists a point c in the interval (a, b) such that $f'(c) = 0$.

Notice that this theorem does not include instructions for how to find that value c. For *theoretical* mathematicians, it is enough to know that such a point exists. *Applied* mathematicians are charged with the task of actually developing strategies for finding the point c. Notice that the theorem does not limit the number of points c in the interval (a, b) that satisfy $f'(c) = 0$. There could be one, five, or twelve thousand. All this theorem states is that there is at least one such number c that satisfies the condition $f'(c) = 0$.

A **critical point** of a function is a point in the domain of the function where the derivative is either 0 or undefined. Rolle's Theorem helps locate some of the critical values of a function. Conceptually, Rolle's Theorem applies to functions that start and end at the same height. If a function initially increases, then it must, at some point, change direction and start to decrease. The point where the function stops increasing and starts decreasing will be a critical point of the function. A similar argument holds if the function initially decreases. At some point the function will have to stop decreasing and start increasing. The point where the function stops decreasing and starts increasing is a critical point of the function. Rolle's Theorem can be used to draw conclusions about functions, as we will see in Example 1.

Example 1

Prove that the polynomial $f(x) = x^3 + x - 1$ has exactly one real root.

Solution: First of all, because $f(x)$ is a polynomial with odd degree and real coefficients, we know that $f(x)$ has at least one real root. Remember that a polynomial of degree n has exactly n roots, but the roots may not be distinct or real. Also, for polynomials with real coefficients, the complex roots travel in conjugate pairs, so that there will always be an even number of complex roots. So a degree 3 polynomial has exactly three roots, and either two of the roots or none of the roots are complex. If two of the roots are complex, then there will be one real root. If none of the roots are complex, then there will be three real roots. So, no matter how you slice it, $f(x) = x^3 + x - 1$ will have either one real root or three real roots. One way to prove that $f(x) = x^3 + x - 1$ has only one real root is to test every real number. Because there are infinitely many real numbers to check, this is not a particularly efficient method. Another way to show that $f(x) = x^3 + x - 1$ has exactly one real root is to see what happens if we assume that $f(x) = x^3 + x - 1$ has *two* real roots. Suppose that there are two values of x, say a and b, that are roots of $f(x) = x^3 + x - 1$. In other words, a and b are distinct real numbers satisfying $f(a) = 0$ and $f(b) = 0$. Polynomials are continuous and differentiable everywhere, so all of the conditions required for Rolle's Theorem are met. Therefore, according to Rolle's Theorem, there is a point c between a and b satisfying $f'(c) = 0$. Well, $f'(x) = 3x^2 + 1$; this is a quadratic function that has no real zeros! There is no point c for which $f'(c) = 0$ for the function $f(x) = x^3 + x - 1$. This is a problem. Rolle's Theorem is valid, meaning that any time the conditions for the theorem are met, the conclusion has to hold. Because the conclusion does not hold, there must be a problem with the conditions. Polynomials are continuous and differentiable, so there is no problem there. Our only choice is to go back to our assumption: that $f(x) = x^3 + x - 1$ has two real roots. That is where we went wrong. If this assumption cannot be true, then $f(x) = x^3 + x - 1$ must only have one real root. And that was what we wanted to prove!

In this previous argument, I never suggested where to look for the one real root. The *Intermediate Value Theorem* can help us locate the root. For the function $f(x) = x^3 + x - 1$, notice that $f(0) = -1$ and $f(1) = 1$. The function changes sign somewhere between $x = 0$ and $x = 1$. The point at which the function changes sign (goes from negative to positive) is the location of the one real root.

Rolle's Theorem is actually a special case of the Mean Value Theorem, and Rolle's Theorem is used to prove the Mean Value Theorem.

The Mean Value Theorem

If a function $f(x)$ is continuous on a closed interval $[a, b]$ and is differentiable on the open interval (a, b), then there is a point c in the interval (a, b) such that $f'(c) = \frac{f(b) - f(a)}{b - a}$.

This theorem does not discuss *how* to find the point c. It merely guarantees that this point will exist. It does not claim that this point is unique. There could be one, three, or two thousand such points in (a, b).

To understand the Mean Value Theorem, consider the situation in which a car is being driven on a stretch of road 140 miles long. If it takes 2 hours to drive that distance, then the average speed is 70 mph. The Mean Value Theorem claims that (assuming the travels were done continuously, without teleportation or any other discontinuous motion) there must be some point during the trip that the car actually traveled at an instantaneous speed of 70 mph. Using cruise control, the car could have traveled 70 mph throughout the entire trip. Another possibility is that the car traveled at a rate of 65 mph for some of the time and 75 mph for part of the time. In this case, the car must have made a transition between 65 mph and 75 mph, and in the process a rate of 70 mph must have been achieved. At some point, the instantaneous velocity must have the same value as the average velocity. That's all that the Mean Value Theorem is trying to say.

Example 2

Does there exist a continuous and differentiable function $f(x)$ such that $f(0) = -1$, $f(2) = 4$, and $f'(x) \leq 2$ for all x?

Solution: Such a function, if it exists, must satisfy the Mean Value Theorem. In other words, there would have to be a point c between $x = 0$ and $x = 2$ such that:

$$f'(c) = \frac{f(2)-f(0)}{2-0} = \frac{4-(-1)}{2-0} = 2.5$$

We are told that $f'(x) \leq 2$ for all x, so the existence of the point c would be problematic. Therefore, no such function could exist.

Lesson 11-4 Review

1. Let $f(x) = \frac{x+1}{x-1}$. Show that there is no value of c satisfying

   $f'(c) = \frac{f(2)-f(0)}{2-0}$. Does this contradict the Mean Value Theorem?

Answer Key

Lesson 11-1 Review

1. The point is (0, 0), $f'(x) = e^x \tan^{-1} x + e^x \left(\frac{1}{1+x^2}\right)$, so the slope is $f'(0) = 1$.

   The equation of the tangent line is $y = x$, $f(0.1) \approx 0.1$, and

   $f(0.1) = e^{0.1} \tan^{-1} 0.1 = 0.1102$.

2. The equation of the tangent line is $y = 4x - 14$, and $f(2.85) \approx -2.6$.

3. The point is (4, G(4)), or $\left(4, \frac{1}{5}\right)$, $G'(x) = \frac{\left(e^{x-4}\right)f(x) - \left(e^{x-4}\right)f'(x)}{\left(f(x)\right)^2}$, so the slope is

   $G'(4) = \frac{8}{25}$. The equation of the tangent line is $y = \frac{8}{25}x - \frac{27}{25}$.

4. The point is (2, G(2)), or (2, 0), $G'(x) = (1)\ln\left(f(x) - 2\right) + x\left(\frac{1}{f(x)-2}\right)\left(f'(x)\right)$,

   so the slope is $G'(2) = -4$. The equation of the tangent line is $y = -4x + 8$, and

   $G(2.2) \approx -0.8$.

5. $f(0) = 1$, $f'(x) = -4(1+2x)^{-5}(2)$, $f'(0) = -8$. The local linearization is $y = -8x + 1$.

Lesson 11-3 Review

1. $V = x^3, dV = 3x^2dx, x = 50, dx = 0.1,\ dV = 3(50)^2(.1) = 750\text{cm}^3$

   $S = 6x^2,\ dS = 12xdx,\ dS = 12(50)(0.1) = 60\text{cm}^2$

2. $y = \sqrt[5]{x},\ dy = \frac{1}{5}x^{-4/5}dx, x = 32, dx = 2,\ dy = \frac{1}{5}(32)^{-4/5}(2) = \frac{1}{40},$

   $\sqrt[5]{34} \approx \sqrt[5]{32} + dy = 2.025$

Lesson 11-4 Review

1. $f'(c) = \frac{f(2)-f(0)}{2-0} = \frac{3-(-1)}{2-0} = 2 > 0$, but $f'(x) = \frac{(1)(x-1)-(x+1)(1)}{(x-1)^2} = \frac{-2}{(x-1)^2} < 0$.

   The derivative is always negative, so there will be no value c for which the derivative is positive. The function $f(x) = \frac{x+1}{x-1}$ is not continuous on [0, 2], so the Mean Value Theorem does not apply.

Further Applications of the Derivative

The chain rule is an important concept in calculus because it enables us to quantify how a change in one variable causes a ripple effect in other quantities. If polluted water trickles into a stream that then feeds a river and goes to the ocean, a change in the concentration of pollutants dumped into the stream will affect the purity of the water in the ocean. This effect can be quantified using the chain rule, and we will discuss these types of problems in the context of related rates.

Another important application of calculus is in optimization. Optimization involves determining the value (or values) of the independent variable that result in the greatest or lowest values in the dependent variable. Our goal could be to maximize our revenue for a business venture, or it could involve minimizing the pollution in our drinking water. We can use calculus to analyze certain situations and help determine optimal solutions.

Lesson 12-1: Related Rates

Related rates problems take into consideration how a change in one quantity results in a change in another quantity. For example, when blowing up a balloon, the surface area of the balloon is changing, and so is the volume of the balloon. The rate at which the volume of the balloon changes will determine the rate at which the surface area of the balloon is changing. The two rates are *related*. In working out these problems, the chain rule is essential.

Here is a basic problem-solving strategy that may help when setting up related rates problems. Related rates problems usually involve functions that are changing with respect to time. It will be important to identify the independent variable and write down appropriate formulas from geometry and trigonometry.

- Read the problem carefully.
- Draw a picture.
- Identify the variables used to represent the quantities under consideration.
- Interpret the rate information given as a derivative.
- Write an equation that relates the various quantities involved in the problem. Use geometry or trigonometry to relate some of the variables to each other to help eliminate them from the equation.
- Differentiate the equation with respect to time using the chain rule and solve for the appropriate quantity.
- Re-read the question to make sure that you have answered it completely.

Example 1

The radius of a spherical balloon is increasing at the rate of $3\frac{\text{cm}}{\text{min}}$. How fast is the volume changing when the radius is 8 cm?

Solution: Let V represent the volume of a sphere, and r represent the radius. Then the volume of the sphere can be calculated using the formula $V = \frac{4}{3}\pi r^3$. We are given that the radius is increasing at a rate of $3\frac{\text{cm}}{\text{min}}$, which means that $\frac{dr}{dt} = 3$. We are asked to find how fast the volume is changing, or $\frac{dV}{dt}$, when $r = 8$. Differentiating the function $V = \frac{4}{3}\pi r^3$ using the chain rule, we have $\frac{dV}{dt} = \frac{4}{3}\pi\left(3r^2\frac{dr}{dt}\right)$. Substituting in for r and $\frac{dr}{dt}$, we have $\frac{dV}{dt} = \frac{4}{3}\pi\left(3(8)^2(3)\right) = 768\pi$. So the volume is changing at a rate of $768\pi\ \frac{\text{cm}^3}{\text{min}}$.

Example 2

A water tank has the shape of an inverted circular cone with base radius 3 m and height 5 m. If water is being pumped into the tank at a rate of $3\frac{\text{m}^3}{\text{min}}$, find the rate at which the water level is rising when the water is 3 m deep.

Solution: First, let's draw a picture. Figure 12.1 shows an inverted cone with some water in it.

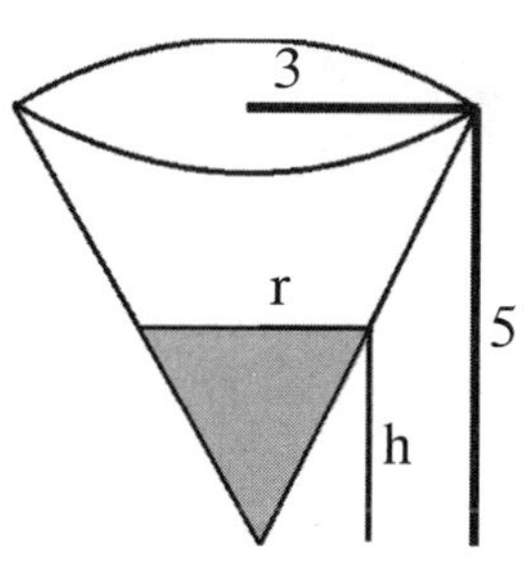

Figure 12.1.

Identify our variables: Let V, r, and h represent the volume, the radius of the water surface, and the height of the water in the cone at time t. Water is being pumped in at a rate of $3\frac{\text{m}^3}{\text{min}}$, meaning $\frac{dV}{dt}=3$, and we need to find $\frac{dh}{dt}$ when $h = 3$. The volume of a cone is given by $V=\frac{1}{3}\pi r^2 h$. As the water is being pumped into the cone, both the height and the radius of the water surface are changing. We need to write the volume of the cone in terms of the height by eliminating r in the equation for the volume. We can use the properties of similar triangles to relate r and h:

$$\frac{r}{h}=\frac{3}{5}$$

Using this equation, we have:

$$r=\frac{3h}{5}$$

Our equation for the volume then becomes:

$$V=\frac{1}{3}\pi\left(\frac{3h}{5}\right)^2 h$$

$$V=\frac{3}{25}\pi h^3$$

Differentiate this function and solve for $\frac{dh}{dt}$ when $\frac{dV}{dt}=3$ and $h = 3$:

$$V=\frac{3}{25}\pi h^3$$

$$\frac{dV}{dt}=\frac{3}{25}\pi\left(3h^2\frac{dh}{dt}\right)$$

$$3=\frac{3}{25}\pi\left(3(3)^2\frac{dh}{dt}\right)$$

$$3 = \frac{81}{25}\pi\frac{dh}{dt}$$

$$\frac{dh}{dt} = \frac{75}{81\pi} = \frac{25}{27\pi} \approx 0.295$$

The height is increasing at a rate of $0.295\frac{\text{m}}{\text{min}}$.

Example 3

An oil tanker is leaking oil that forms a circular oil slick that is 0.1 ft thick. To estimate the rate at which the oil is leaking from the tanker, it was found that the radius of the slick was increasing at a rate of $0.32\frac{\text{ft}}{\text{min}}$ when the radius was 500 ft. Find the rate at which the oil is leaking from the tanker.

Solution: The oil slick is cylindrical in shape, with a radius of 500 ft and a thickness of 0.1 ft. Let V, r, and h represent the volume, the radius of the oil slick, and the thickness of the oil slick at time t. We are given that $\frac{dr}{dt} = 0.32$ and $h = 0.1$, and we are asked to find $\frac{dV}{dt}$. The volume of a cylinder can be calculated using the formula $V = \pi r^2 h$. Differentiating this equation with respect to time, and assuming that the thickness of the oil slick is constant, we have:

$$V = \pi r^2 h$$

$$\frac{dV}{dt} = \pi\left(2r\frac{dr}{dt}\right)h$$

Substituting in for r, h, and $\frac{dr}{dt}$, we have: $$\frac{dV}{dt} = \pi(2(500)(0.32))(0.1)$$

$$\frac{dV}{dt} = 32\pi \approx 100.53$$

The oil is leaking out of the tanker at a rate of approximately $100.53\frac{\text{ft}^3}{\text{min}}$.

Lesson 12-2: L'Hopital's Rule

When we first introduced the idea of a limit, we ran across limits of an indeterminate form, or limits of the form $\frac{0}{0}$ or $\frac{\infty}{\infty}$. At that time, we had not discussed the derivative, and we had to use algebraic simplification techniques to evaluate these limits. As it turns out, the derivative can be used to evaluate some limits of this type, and this technique is referred to as L'Hopital's Rule. Just as we saw with the Intermediate Value Theorem, Rolle's Theorem, and the Mean Value Theorem, we must be sure that the criteria for applying L'Hopital's Rule are met before applying it.

L'Hopital's Rule

Suppose $f(x)$ and $g(x)$ are differentiable functions and there exists an open interval I containing a point a such that $g'(x) \neq 0$ for all values x (except possibly at a) contained in I.

Suppose also that either $\lim_{x\to a} f(x) = \lim_{x\to a} g(x) = 0$

or $\lim_{x\to a} f(x) = \lim_{x\to a} g(x) = \pm\infty$.

Then $\lim_{x\to a} \frac{f(x)}{g(x)} = \lim_{x\to a} \frac{f'(x)}{g'(x)}$, if $\lim_{x\to a} \frac{f'(x)}{g'(x)}$ exists.

If we are evaluating a limit of a quotient, and the numerator and the denominator either *both* approach 0 or *both* approach $\pm\infty$, then we can take the derivative of the numerator and the derivative of the denominator separately (not the derivative of the quotient, but differentiate each individual function separately), and evaluate the limit of the ratio of the derivatives. L'Hopital's Rule will enable us to evaluate some of the limits that gave us trouble in Chapter 5, and will also enable us to evaluate some of the limits that we had to use numerical methods to evaluate.

Example 1

Evaluate the following limits using L'Hopital's Rule:

a. $\lim_{x\to 0} \frac{e^{2x} + x - 1}{x}$

b. $\lim_{x\to 0} \frac{2\tan x - x}{3x}$

c. $\lim_{x\to 1} \frac{\ln x}{x - 1}$

Solution: For each of these problems, first make sure that the conditions for applying L'Hopital's Rule are met. If they are, then apply the rule:

a. $\lim_{x \to 0} \frac{e^{2x}+x-1}{x}$: Both the numerator and the denominator approach 0 as $x \to 0$, so we can apply L'Hopital's Rule. Differentiate both the numerator and the denominator separately, and try to take the limit again:

$$\lim_{x \to 0} \frac{e^{2x}+x-1}{x} = \lim_{x \to 0} \frac{2e^{2x}+1}{1} = 3$$

b. $\lim_{x \to 0} \frac{2\tan x - x}{3x}$: Both the numerator and the denominator approach 0 as $x \to 0$, so we can apply L'Hopital's Rule. Differentiate both the numerator and the denominator separately, and try to take the limit again:

$$\lim_{x \to 0} \frac{2\tan x - x}{3x} = \lim_{x \to 0} \frac{2\sec^2 x - 1}{3} = \frac{1}{3}$$

c. $\lim_{x \to 1} \frac{\ln x}{x-1}$: Both the numerator and the denominator approach 0 as $x \to 1$, so we can apply L'Hopital's Rule. Differentiate both the numerator and the denominator separately, and try to take the limit again:

$$\lim_{x \to 1} \frac{\ln x}{x-1} = \lim_{x \to 1} \frac{1/x}{1} = 1$$

There are times when, after taking the derivative of the numerator and the denominator, we are still left with an indeterminate form. If the conditions for L'Hopital's Rule continue to be met, the rule can be applied again.

Example 2

Evaluate the following limits using L'Hopital's Rule:

a. $\lim_{x \to 0} \frac{e^x - x - 1}{x^2}$

b. $\lim_{x \to 0} \frac{\tan x - x}{3x^2}$

Solution: For each of these problems, first make sure that the conditions for applying L'Hopital's Rule are met. If they are, then apply the rule repeatedly until you can take the limit.

a. $\lim_{x\to 0}\frac{e^x-x-1}{x^2}$: Both the numerator and the denominator approach 0 as $x\to 1$, so we can apply L'Hopital's Rule. Differentiate both the numerator and the denominator separately, and try to take the limit again:

$$\lim_{x\to 0}\frac{e^x-x-1}{x^2}=\lim_{x\to 0}\frac{e^x-1}{2x}=\lim_{x\to 0}\frac{e^x}{2}=\frac{1}{2}$$

b. $\lim_{x\to 0}\frac{\tan x-x}{3x^2}$: Both the numerator and the denominator approach 0 as $x\to 1$, so we can apply L'Hopital's Rule. Differentiate both the numerator and the denominator separately, and try to take the limit again:

$$\lim_{x\to 0}\frac{\tan x-x}{3x^2}=\lim_{x\to 0}\frac{\sec^2 x-1}{6x}=\lim_{x\to 0}\frac{2\sec x(\sec x\tan x)}{6}=0$$

Sometimes a problem is written in a form in which it seems as if the criteria for using L'Hopital's Rule are not met, but after a little rearrangement L'Hopital's Rule does apply.

Example 3

Evaluate $\lim_{x\to 0} x\ln x$.

Solution: In this situation, we are taking the limit of the product of two functions. The first function is heading towards 0 while the second function is heading towards $-\infty$. The question is: Which function wins? This limit is an indeterminate of the form $0\cdot\infty$, and L'Hopital's Rule does not apply, but if we rewrite the problem as $\lim_{x\to 0}\frac{\ln x}{1/x}$, we have an indeterminate of the form $\frac{\infty}{\infty}$, and L'Hopital's Rule does apply:

$$\lim_{x\to 0} x\ln x=\lim_{x\to 0}\frac{\ln x}{1/x}=\lim_{x\to 0}\frac{1/x}{-1/x^2}=\lim_{x\to 0}1/x\cdot\left(-x^2\right)=\lim_{x\to 0}(-x)=0$$

Some interesting complications arise when evaluating limits of the form $\lim_{x\to a}\left[f(x)\right]^{g(x)}$:

- If $\lim_{x\to a} f(x) = \lim_{x\to a} g(x) = 0$, we have an indeterminate form of the type 0^0.
- If $\lim_{x\to a} f(x) = \infty$ and $\lim_{x\to a} g(x) = 0$, we have an indeterminate form of the type ∞^0.
- If $\lim_{x\to a} f(x) = 1$ and $\lim_{x\to a} g(x) = \pm\infty$, we have an indeterminate form of the type 1^∞.

When evaluating limits of the type $\lim_{x\to a}[f(x)]^{g(x)}$, use the rules for exponents to write the problem in a different format:

$$\lim_{x\to a}[f(x)]^{g(x)} = \lim_{x\to a}\left[e^{\ln f(x)}\right]^{g(x)} = \lim_{x\to a} e^{g(x)\cdot\ln f(x)}$$

Now try to evaluate $\lim_{x\to a} g(x)\cdot\ln f(x)$ using L'Hopital's Rule (if necessary) and then use the fact that $\lim_{x\to a} e^{g(x)\cdot\ln f(x)} = e^{\lim_{x\to a} g(x)\cdot\ln f(x)}$ to finish the problem.

Example 4

Evaluate the limit: $\lim_{x\to\infty}\left(\frac{x}{x+1}\right)^x$

Solution: Because $\lim_{x\to\infty}\left(\frac{x}{x+1}\right)^x$ is an indeterminate of the form 1^∞, we need to rewrite the limit as $\lim_{x\to\infty}\left(\frac{x}{x+1}\right)^x = \lim_{x\to\infty} e^{x\cdot\left(\ln\left(\frac{x}{x+1}\right)\right)}$, then evaluate $\lim_{x\to\infty} x\cdot\left(\ln\left(\frac{x}{x+1}\right)\right)$:

Rewrite the problem so that L'Hopital's Rule applies:

$$\lim_{x\to\infty} x\cdot\left(\ln\left(\frac{x}{x+1}\right)\right) = \lim_{x\to\infty}\frac{\ln x - \ln(x+1)}{1/x}$$

Apply L'Hopital's Rule $\qquad \lim_{x\to\infty}\frac{\ln x - \ln(x+1)}{1/x} = \lim_{x\to\infty}\frac{1/x - 1/(x+1)}{-1/x^2}$

Invert the denominator and multiply

$$\lim_{x\to\infty}\frac{\frac{1}{x}-\frac{1}{x+1}}{-\frac{1}{x^2}}=\lim_{x\to\infty}\left(\tfrac{1}{x}-\tfrac{1}{x+1}\right)\cdot\left(-x^2\right)$$

Get a common denominator and subtract $\frac{1}{x+1}$ from $\frac{1}{x}$

$$\lim_{x\to\infty}\left(\tfrac{1}{x}-\tfrac{1}{x+1}\right)\cdot\left(-x^2\right)=\lim_{x\to\infty}\left(\frac{1}{x(x+1)}\right)\cdot\left(-x^2\right)$$

Multiply the two terms together and expand the denominator

$$\lim_{x\to\infty}\left(\frac{1}{x(x+1)}\right)\cdot\left(-x^2\right)=\lim_{x\to\infty}\frac{-x^2}{x^2+x}$$

Evaluate the limit $\quad\lim_{x\to\infty}\frac{-x^2}{x^2+x}=-1$

Finally, evaluate the original limit

$$\lim_{x\to\infty}\left(\tfrac{x}{x+1}\right)^x=\lim_{x\to\infty}e^{x\cdot\left(\ln\left(\frac{x}{x+1}\right)\right)}=e^{-1}$$

Lesson 12-2 Review

Evaluate the following limits:

1. $\lim_{x\to\infty}\frac{\ln(\ln x)}{\sqrt{x}}$

2. $\lim_{x\to 0}\frac{\tan^{-1}2x}{3x}$

3. $\lim_{x\to -\infty} xe^x$

4. $\lim_{x\to 0}\frac{\sin x}{\sinh x}$

5. $\lim_{x\to\infty}\left(1+\tfrac{1}{x^2}\right)^x$

Lesson 12-3: The Extreme Value Theorem and Optimization

There are many situations in which determining the largest or smallest value of some quantity is important. For example, businesses typically want to maximize profit, and people concerned with the environment try to minimize resource consumption. An **optimization** problem is a problem that involves finding the optimal, or best, way of doing something. In many situations, these problems can be turned into finding the maximum or minimum value of a function, and calculus can be used to solve these types of problems.

A function $f(x)$ has an **absolute maximum**, or **global maximum**, at $x = c$ if $f(c) \geq f(x)$ for all x in the domain of $f(x)$. The value $f(c)$ is called the absolute maximum value of $f(x)$ on its domain. We can define the absolute minimum in a similar manner. A function $f(x)$ has an **absolute minimum**, or **global minimum**, at $x = c$ if $f(c) \leq f(x)$ for all x in the domain of $f(x)$. The value $f(c)$ is called the absolute minimum value of $f(x)$ on its domain. Notice that these definitions do not involve any ideas from calculus. The global maximum is the highest point of the graph of $f(x)$, and the global minimum is the lowest point of the graph of $f(x)$. Not all functions have a global maximum or a global minimum. Figure 12.2 shows graphs of functions, some with global maxima or minima and some without.

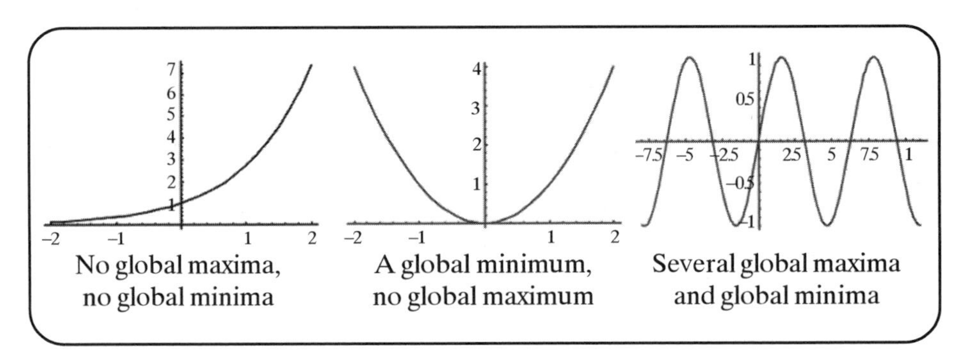

Figure 12.2.

We can use calculus to help find global maxima and global minima. The Extreme Value Theorem gives conditions under which a function has a global maxima and a global minima. This is another example of an existence theorem in mathematics: We know when a function is guaranteed to have a global maximum and a global minimum, but the theorem does not tell us how to actually find these special values. Fortunately, we also have a set of instructions for how to find the global maximum and global minimum when the criteria for the Extreme Value Theorem are met.

Extreme Value Theorem

If $f(x)$ is continuous on a closed interval $[a, b]$, then there exist points c and d in $[a, b]$ such that $f(c)$ is an absolute maximum and $f(d)$ is an absolute minimum.

As with all theorems, you must be careful when you apply it. This theorem does not say that the absolute maximum and minimum are unique. There could be several places where the absolute maximum and minimum are attained. For example, the function $f(x) = \sin x$ has a maximum value of 1 and a minimum value of –1, and on the closed interval $[-10\pi, 10\pi]$ there are several locations where the function attains its maximum and minimum values. The Extreme Value Theorem merely guarantees that there is a global maximum and a global minimum and that the function does, in fact, attain those values somewhere in the given interval.

The closed interval in the Extreme Value Theorem must have finite length. To understand why, consider the function $f(x) = e^{-x}$ over the closed interval $[0, \infty)$. This function has a horizontal asymptote at $y = 0$, but there is no value of x for which e^{-x} is 0; $f(x) = e^{-x}$ is bounded below by $y = 0$, but it never reaches this lower bound.

Continuity is also very important in this theorem. Any break in the curve may result in the function never attaining its absolute maximum or absolute minimum values. If the criteria for the Extreme Value Theorem are met, then the function is guaranteed to attain its absolute maximum and absolute minimum values. If these values exist, it is a simple matter of developing an algorithm for finding those values. Fortunately, such an algorithm exists, and it involves calculus.

The following procedure can be used to find the absolute maximum and absolute minimum values of a continuous function $f(x)$ on a closed interval $[a, b]$:

- Find the critical values of $f(x)$ in the closed interval $[a, b]$. Remember that the critical values of a function are the points in the domain where the derivative is either 0 or undefined. Suppose $f(x)$ has n critical values in $[a, b]$: $c_1, c_2 \ldots, c_n$.
- Evaluate the function at each of the critical values found in the previous step. Also, evaluate the function at the endpoints of the interval $[a, b]$. In other words, evaluate $f(a), f(b), f(c_1), f(c_2), \ldots, f(c_n)$.

- The largest of the values $f(a)$, $f(b)$, $f(c_1)$, $f(c_2)$, ..., $f(c_n)$ is the *absolute maximum* of $f(x)$, and the smallest of the values $f(a)$, $f(b)$, $f(c_1)$, $f(c_2)$, ..., $f(c_n)$ is the *absolute minimum* of $f(x)$.

Example 1

Find the absolute maximum and the absolute minimum values of the function $f(x) = x^3 - 3x^2 + 3$ over the interval $[1, 4]$.

Solution: Follow the steps outlined previously.

Step 1: Find the critical values of $f(x) = x^3 - 3x^2 + 3$ in $[1, 4]$ by solving the equation $f'(x) = 0$:

Take the derivative	$f'(x) = 3x^2 - 6x$
Set the derivative equal to 0	$3x^2 - 6x = 0$
Factor	$3x(x-2) = 0$
Set each factor equal to 0	$x = 0$ or $x = 2$

The critical values of $f(x) = x^3 - 3x^2 + 3$ are $x = 0$ or $x = 2$, but only one of the critical values, $x = 2$, is in our interval. The critical value $x = 0$ is not in our interval, so we will not include it in the next step.

Step 2: Evaluate the function at the critical value ($x = 2$) and at the endpoints of the interval ($x = 1$ and $x = 4$):

$$f(2) = -1, f(1) = 1, \text{ and } f(3) = 3$$

Step 3: Identify the absolute maximum and the absolute minimum. From this list, we see that the absolute maximum value is $f(3) = 3$ and the absolute minimum value is $f(2) = -1$. Figure 12.3 shows a graph of $f(x) = x^3 - 3x^2 + 3$ over the interval $[1, 4]$.

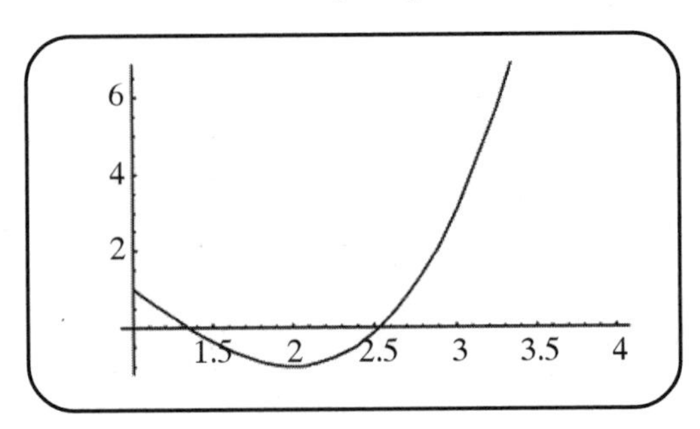

Figure 12.3.

A local maximum is similar to a big fish in a little pond. Officially, a function $f(x)$ has a **local maximum**, or **relative maximum**, at $x = c$ if there is an open interval I containing c such that $f(c) \geq f(x)$ for all x in I. Similarly, a function $f(x)$ has a **local minimum**, or **relative minimum**, at $x = c$ if there is an open interval I containing c such that $f(c) \leq f(x)$ for all x in I. The difference between a *global* maximum and a *local* maximum is that the inequality $f(c) \geq f(x)$ holds for all values of x in the domain for a *global* maximum, whereas for a *local* maximum, the inequality $f(c) \geq f(x)$ only has to hold for values of x in an open interval around c. Fortunately, calculus can help locate these local maxima and local minima as well.

Fermat's Theorem

If $f(x)$ has a local maximum or local minimum at $x = c$,
and if $f'(c)$ exists, then $f'(c) = 0$.

This is an important characterization of local maxima and local minima. If a function is differentiable on its domain, and if the function has any local maxima or minima, they can only be located at places where the derivative is 0. In fact, if a function has *any* local maxima and minima, the best place to start looking for them are at the critical values of the function. Fermat's Theorem does *not* say that $f'(c) = 0$ implies that $x = c$ is a local maximum or a local minimum. It says that *if* there is a local maximum or a local minimum, and *if* the derivative of the function exists at that local maximum or local minimum, *then* the derivative must be 0. A function *can* have critical values that are neither local maxima nor local minima.

Optimization problems can show up directly or indirectly, as we will see in the next two examples.

Example 2

Show that $x > 2 \ln x$ for all $x > 0$.

Solution: Demonstrating the inequality $x > 2 \ln x$ is equivalent to showing that $x - 2 \ln x > 0$. This inequality will be true if the function $f(x) = x - 2 \ln x$ is always positive. We can show this by showing that the minimum value of the function $f(x) = x - 2 \ln x$ is a positive number. If the minimum value of $f(x) = x - 2 \ln x$ is positive, then all other values of the function $f(x) = x - 2 \ln x$ will be positive, so the inequality will be established. To minimize the function $f(x) = x - 2 \ln x$, look for critical values:

Differentiate the function:

$$f'(x) = 1 - \frac{2}{x}$$

Set the derivative equal to 0 to find the critical values:

$$1 - \frac{2}{x} = 0$$
$$\frac{2}{x} = 1$$
$$x = 2$$

The critical value of the function is at $x = 2$, and $f(2) = 2 - 2\ln 2 = 0.6137 > 0$. The graph of $f(x) = x - 2\ln x$ is shown in Figure 12.4. Notice that for values of x in the interval $(0, 2)$, $f(x)$ is decreasing, and in the interval $(2, \infty)$ the function $f(x)$ is increasing. It reaches its minimum at its critical value, when $x = 2$.

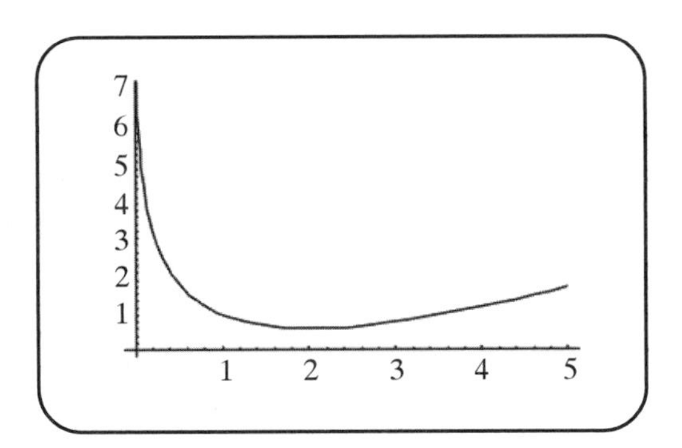

Figure 12.4.

Inequalities can often be established by creating a function and either finding the maximum or minimum of that function. If the maximum or minimum value of the function satisfies the inequality, there is a good chance that the inequality is true.

Example 3

An aluminum can contains 0.355 cm^3 of soda. Find the dimensions of the can that use the least amount of material. Assume that the can is cylindrical and is capped at both ends.

Solution: It is possible for the aluminum can to be shaped as a straw: long and narrow. In this case, the top and bottom do not require much material, but the sides of the can require a lot of material. It is also possible for the aluminum can to be shaped similar to a can of cat food: short and squatty. In that case, the sides of the can do not require much material, but the top and bottom will. There's a trade-off: The more narrow the can, the more material needed for the sides, and the shorter the can, the more material needed for the top and bottom. We will want to find the balance. First, write an equation for the total surface area of the aluminum can. The surface area of a cylinder of radius r and height h is $2\pi rh$. The surface area of the top and bottom is $2 \cdot (\pi r^2)$. The total surface area of the aluminum can is $2\pi rh + 2\pi r^2$. There are two variables in this equation. We need to eliminate one of them by using the constraint that the volume of the can is 0.355 cm³. The volume of a cylinder is given by the equation $V = \pi r^2 h$, so we have $\pi r^2 h = 0.355$, or $h = \frac{0.355}{\pi r^2}$. We can substitute this equation for h into the formula for the surface area to get:

$$S(r) = 2\pi rh + 2\pi r^2 = 2\cancel{\pi} r\left(\frac{0.355}{\cancel{\pi} r^2}\right) + 2\pi r^2 = \left(\frac{0.710}{r}\right) + 2\pi r^2$$

We can look at each piece of the surface area equation. The first part of the equation, $\left(\frac{0.710}{r}\right)$, is the contribution from the side of the can. As the radius decreases, so that the can takes on the shape of a straw, most of the surface area comes from the sides and $\left(\frac{0.710}{r}\right)$ becomes large. The second term in the surface area of the can, $2\pi r^2$, quantifies the contribution of the top and bottom to the surface area. As the radius increases, the can becomes short and squatty, and the surface area of the top and bottom contribute the most to the overall surface area. There will be a radius that provides a balance, or a minimum surface area. To find this radius, differentiate the surface area equation and look for the critical values:

Differentiate the surface area function $\quad S(r) = \left(\frac{0.710}{r}\right) + 2\pi r^2$

Set $S'(r)$ equal to 0 and solve for r $\quad S'(r) = \left(\frac{-0.710}{r^2}\right) + 4\pi r$

$$\left(\frac{-0.710}{r^2}\right) + 4\pi r = 0$$

$$4\pi r = \frac{0.710}{r^2}$$

$$r^3 = \frac{0.710}{4\pi}$$

$$r = \sqrt[3]{\frac{0.710}{4\pi}} \approx 0.384$$

Now make sure that we answer the question. The units for volume were given as cm^3, so our radius has units of cm. The radius that minimizes the surface area of our can is 0.384 cm. The graph of the surface area of the can as a function of the radius is shown in Figure 12.5.

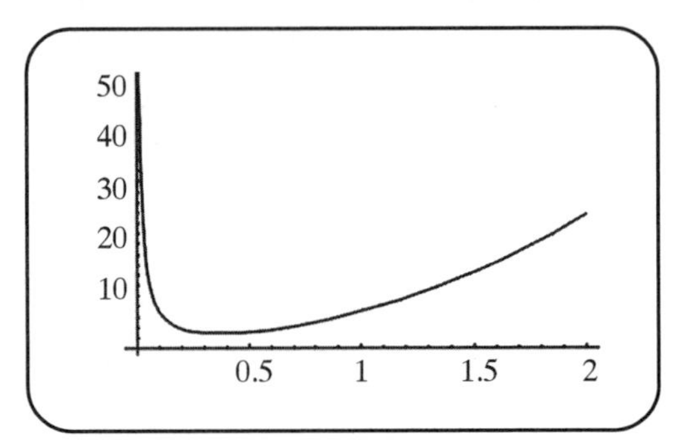

Figure 12.5.

Answer Key

Lesson 12-2 Review

1. $\lim\limits_{x\to\infty} \frac{\ln(\ln x)}{\sqrt{x}} = \lim\limits_{x\to\infty} \frac{\left(\frac{1}{\ln x}\right)\left(\frac{1}{x}\right)}{\frac{1}{2}x^{-1/2}} = \lim\limits_{x\to\infty} \frac{2}{x^{1/2}\ln x} = 0$

2. $\lim\limits_{x\to 0} \frac{\tan^{-1} 2x}{3x} = \lim\limits_{x\to 0} \frac{\left(\frac{1}{1+(2x)^2}\right)(2)}{3} = \frac{2}{3}$

3. $\lim\limits_{x\to-\infty} xe^x = \lim\limits_{x\to-\infty} \frac{x}{e^{-x}} = \lim\limits_{x\to-\infty} \frac{1}{-e^{-x}} = 0$

4. $\lim\limits_{x\to 0} \frac{\sin x}{\sinh x} = \lim\limits_{x\to 0} \frac{\cos x}{\cosh x} = 1$

5. $\lim\limits_{x\to\infty}\left(1+\frac{1}{x^2}\right)^x=\lim\limits_{x\to\infty}e^{x\ln\left(1+\frac{1}{x^2}\right)}$, and

$$\lim_{x\to\infty}x\ln\left(1+\tfrac{1}{x^2}\right)=\lim_{x\to\infty}\frac{\ln\left(1+\frac{1}{x^2}\right)}{\frac{1}{x}}=\lim_{x\to\infty}\frac{\left(\frac{1}{1+\frac{1}{x^2}}\right)\left(-\frac{2}{x^3}\right)}{-\frac{1}{x^2}}=\lim_{x\to\infty}\frac{\left(\frac{-2}{x^3+x}\right)}{-\frac{1}{x^2}}$$

$$=\lim_{x\to\infty}\left(\frac{-2}{x^3+x}\right)\left(-x^2\right)=\lim_{x\to\infty}\left(\frac{2x^2}{x^3+x}\right)=0$$

and $\lim\limits_{x\to\infty}\left(1+\frac{1}{x^2}\right)^x=\lim\limits_{x\to\infty}e^{x\ln\left(1+\frac{1}{x^2}\right)}=e^0=1$

Graphical Analysis Using the First Derivative

One very important application of the derivative has to do with graphical analysis. The days of plotting points will soon be left behind. Of course, now that graphing calculators are so abundant, the usefulness of calculus in the realm of graphical analysis may be somewhat diminished. I believe that the ability to analyze a function without having to rely on technology is important, especially when the information you have about a function is in terms of its *rate of change*, or its derivative.

Lesson 13-1: The First Derivative

A function $f(x)$ is increasing on an interval I if $f(a) < f(b)$ whenever $a < b$. We can analyze the sign of the difference quotient and see how the derivative of the function comes into play: $\frac{f(b)-f(a)}{b-a} > 0$ and $\lim_{b \to a} \frac{f(b)-f(a)}{b-a} \geq 0$, if this limit exists. In other words, if a function is increasing and differentiable on an interval I, then $f'(x) \geq 0$ on that interval.

Given a differentiable function, the test for monotonic functions uses calculus to determine the intervals where the function is increasing and where the function is decreasing.

Test for Monotonic Functions

Suppose $f(x)$ is continuous on $[a, b]$ and is differentiable on (a, b).
If $f'(x) > 0$ for all x in (a, b), then $f(x)$ is increasing on $[a, b]$.
If $f'(x) < 0$ for all x in (a, b), then $f(x)$ is decreasing on $[a, b]$.

The proof of the test for monotonic functions makes use of the Mean Value Theorem. Our focus will be on applying this test to specific functions.

To determine the regions where a function is increasing or decreasing, we can follow the following procedure:

- Differentiate the function.
- Factor the derivative completely.
- Find all roots (zeros) of the derivative.
- Find all points of discontinuity of the derivative.
- Arrange the roots and the discontinuities on a number line. Because of the Intermediate Value Theorem, these points are the only places where the derivative can change sign. Create a sign chart for the derivative of the function. Choose a test value in each of the regions on the number line and determine the sign of the derivative at those test values.

Once you have created a sign chart, it is easy to determine where the function is increasing and where it is decreasing.

Example 1

Find the regions where the function $f(x) = 3x^4 - 4x^3 - 12x^2 + 8$ is increasing and where it is decreasing.

Solution: Follow the method outlined.

- Differentiate the function: $f'(x) = 12x^3 - 12x^2 - 24x$
- Factor the derivative completely: $f'(x) = 12x(x^2 - x - 2) = 12x(x - 2)(x + 1)$
- Find all roots (zeros) of the derivative: $x = 0$, $x = 2$ and $x = -1$.
- Find all points of discontinuity of the derivative: A polynomial is continuous everywhere, so there are no discontinuities.
- Arrange the roots and the discontinuities on a number line, and create a sign chart for $f'(x)$. The points $x = 0$, $x = 2$, and $x = -1$ break the number line into four regions. The test values and the signs of the derivative are listed in the table shown here.

$x=-2$	$f'(-2)<0$
$x=-\frac{1}{2}$	$f'\left(-\frac{1}{2}\right)>0$
$x=1$	$f'(1)<0$
$x=3$	$f'(3)>0$

The sign chart for $f'(x)$ is shown in Figure 13.1.

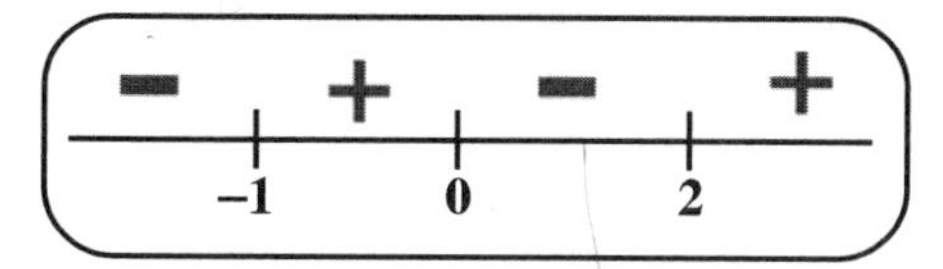

Figure 13.1.

Based on this sign chart, $f(x) = 3x^4 - 4x^3 - 12x^2 + 8$ is increasing in the region $(-1, 0) \cup (2, \infty)$ and is decreasing in the region $(-\infty, -1) \cup (0, 2)$. At the partition points, $x = 0$, $x = 2$, and $x = -1$, the derivative is equal to 0, so something else is happening at those points. Figure 13.2 shows the graph of $f(x) = 3x^4 - 4x^3 - 12x^2 + 8$. Notice that the function is increasing in the region $(-1, 0) \cup (2, \infty)$ and is decreasing in the region $(-\infty, -1) \cup (0, 2)$, as we expected.

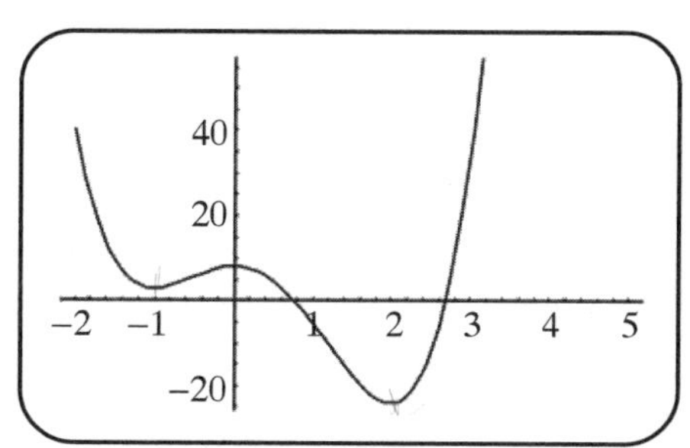

Figure 13.2.

Calculus provides a way of determining the regions where a function is increasing or decreasing. If a differentiable function is increasing, it will have a positive derivative; if it is decreasing, it will have a negative derivative.

Lesson 13-1 Review

Find the regions where the following functions are increasing and where they are decreasing.

1. $g(x) = x\sqrt{1-x^2}$
2. $f(x) = xe^x$

Lesson 13-2: Using the First Derivative

Recall that if a function $f(x)$ has a local maximum or minimum at $x = c$, then by Fermat's Theorem, c must be a critical point. Keep in mind, however, that not all critical points turn out to be local maxima or minima. Calculus will help us determine whether a critical point is a local maximum, a local minimum, or neither. The term **local extrema** refers to either local maxima or local minima.

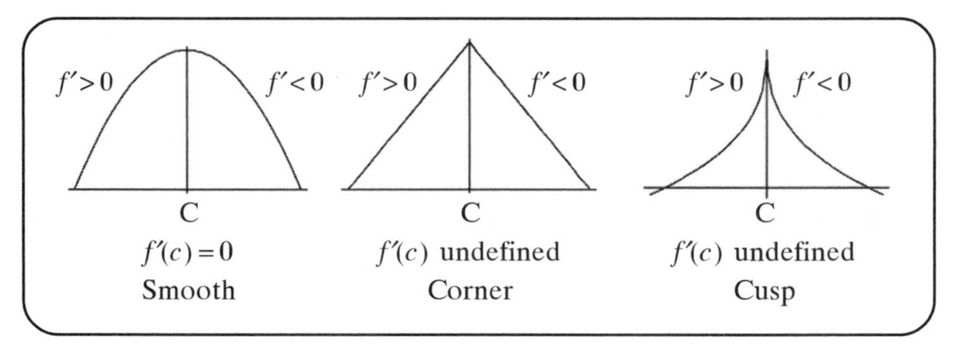

Figure 13.3.

The functions shown in Figure 13.3 all have one thing in common: Each graph has a local maximum. The first graph is smooth (as with a parabola), the second graph has a corner (reminding us of an absolute value function), and the third graph has a cusp. Each of these local maxima occur at critical points: In the first graph, the derivative at the local maximum is 0, and for the other two functions the derivative at the local maximum does not exist. Let's examine the sign of the derivative to the left and the right of each of these critical points. To the left of these critical points, the function is increasing, and its derivative is positive. To the right of these critical points, the function is decreasing, and its derivative is negative. If $x = c$ is a critical point of a function $f(x)$, and the derivative changes from positive to negative around the critical point, then the critical point will be a local maximum of the function. A similar characterization holds for local minima. In general, if the derivative of a function changes sign around a critical point, then that critical point will either result in a local maximum or a local minimum of the function. If the derivative does not change sign around the critical point, then the critical point will not result in a local maximum or a local minimum for the function. The change in sign must accompany the critical value in order for a local maximum or local minimum to result. This can be summarized as the First Derivative Test.

The First Derivative Test

Suppose that $x = c$ is a critical point of a continuous function $f(x)$.
If $f'(x)$ changes sign from positive to negative around $x = c$, then $f(x)$ has a local maximum at $x = c$.
If $f'(x)$ changes sign from negative to positive around $x = c$, then $f(x)$ has a local minimum at $x = c$.
If $f'(x)$ does not change sign around $x = c$, then $f(x)$ has no local extremum at $x = c$.

Example 1

Find the local extrema of $f(x)=x(1-x)^{2/5}$.

Solution: The key to finding local extrema is to examine every critical point of the function. Remember that critical points are points in the domain where the derivative is either 0 or is undefined. First, differentiate the function using the product rule and the chain rule:

Use the product rule $\quad f'(x)=(x)'\left((1-x)^{2/5}\right)+(x)\left((1-x)^{2/5}\right)'$

Use the chain rule to differentiate $(1-x)^{2/5}$

$$f'(x)=(1)\left((1-x)^{2/5}\right)+(x)\left(\tfrac{2}{5}(1-x)^{-3/5}(-1)\right)$$

$$f'(x)=(1-x)^{2/5}-\tfrac{2}{5}x(1-x)^{-3/5}$$

Simplify the derivative to find the critical values (The algebra will get a bit messy, which is why strong algebra skills are so important in calculus)

$$f'(x)=(1-x)^{2/5}-\tfrac{2}{5}x(1-x)^{-3/5}$$

Get a common denominator to combine the two expressions

$$f'(x)=\frac{(1-x)}{(1-x)^{3/5}}-\frac{\frac{2}{5}x}{(1-x)^{3/5}}$$

Combine the numerators $\quad \dfrac{(1-x)-\frac{2}{5}x}{(1-x)^{3/5}}$

Simplify $\quad \dfrac{1-\frac{7}{5}x}{(1-x)^{3/5}}$

The derivative will equal 0 when the *numerator* equals 0: $x = \frac{5}{7}$. Another possible critical point is located where the derivative does not exist, or where the *denominator* equals 0: $x = 1$. Because $x = 1$ is in the domain of the original function, $x = 1$ is also a critical point. Now we need to determine whether the points $x = \frac{5}{7}$ and $x = 1$ are local maxima, local minima, or neither. We can set up a sign chart for the derivative $f'(x) = \frac{1 - \frac{7}{5}x}{(1-x)^{2/5}}$ to determine whether or not the required sign change occurs. (See Figure 13.4.) From this sign chart,

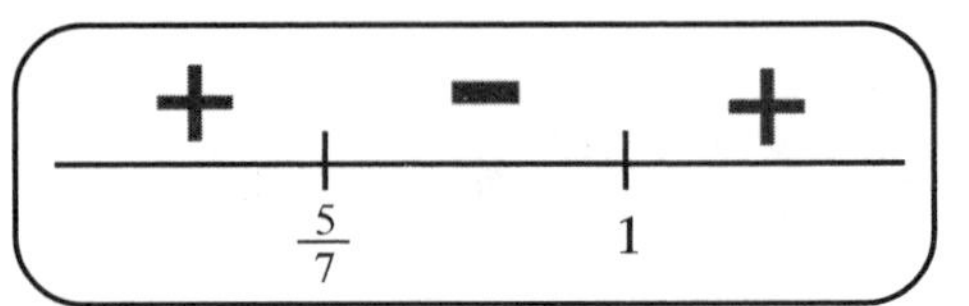

Figure 13.4.

we can conclude that $f(x) = x(1-x)^{3/5}$ has a local maxima at $x = \frac{5}{7}$ and a local minima at $x = 1$. The graph of $f(x) = x(1-x)^{3/5}$ is shown in Figure 13.5. The graph has a cusp at $x = 1$ because the slope of the tangent line at $x = 1$ is not defined at $x = 1$.

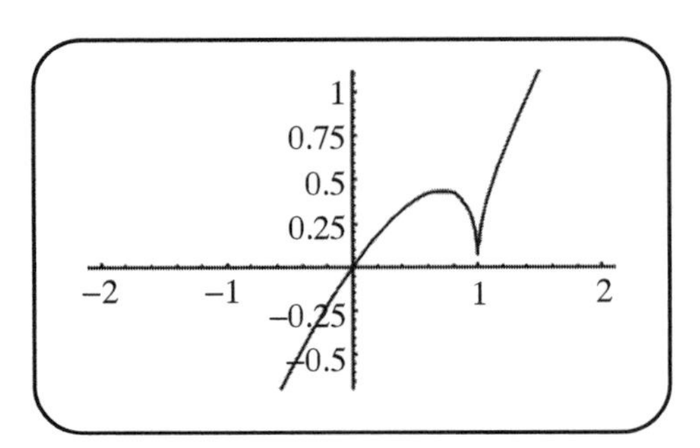

Figure 13.5.

Lesson 13-2 Review

1. Find the local and absolute extreme values of $f(x) = x - 2\sin x$ on $[0, 2\pi]$.

Lesson 13-3: Relating the Graphs of $f(x)$ and $f'(x)$

Given a graph of a function, we can sketch a graph of its derivative. Our graphs will not be precise, but rather they will give a basic idea of the general behavior of the function. Our graphs may not be accurate, but there are some important regions that must be correctly identified in our sketches. Specifically, it is important to identify the regions where the function is increasing and where the function is decreasing. If a function is increasing, its derivative will be positive, and if a function is decreasing, its derivative will be negative. The points where the function stops increasing and starts decreasing, or vice versa, are where the local extrema are located. These points are critical points where the derivative is equal to 0 if the graph is smooth, undefined if the graph has a cusp, or discontinuous (with a jump discontinuity) or kinked if the graph has a corner. Once you have the domain divided into regions where the function is increasing or decreasing, graphing the derivative involves drawing a function that is positive (or lies above the x-axis) in the regions where the original function is increasing and is negative (or lies below the x-axis) in the regions where the original function is decreasing.

At this point in our discussion, if your sketch of the derivative has the correct position relative to the x-axis, you are on the right track. Our graphs will become more precise as we use calculus to analyze our graphs further.

If you need some help visualizing this process, work with a concrete function and its derivative. For example, the derivative of the function $f(x) = x^2$ is $f'(x) = 2x$. We can visualize the graph of $f(x) = x^2$: It is a parabola that is concave up, decreases on the interval $(-\infty, 0)$, increases on the interval $(0, \infty)$, and has a local minimum at $x = 0$. The function $f'(x) = 2x$ is equal to 0 at $x = 0$, is negative on the interval $(-\infty, 0)$, and is positive on the interval $(0, \infty)$. In other words, $f(x)$ is increasing in the region where $f'(x) > 0$, and $f(x)$ is decreasing in the region where $f'(x) < 0$. The local minima occurs where $f'(x) = 0$ and changes sign from negative to positive. If you graph these two functions, you will see the connections between the signs of the derivative and the behavior of the function.

Feel free to analyze the graphs of some other familiar functions, compute the derivative algebraically, and then graph the derivative. A graphing calculator will help you in this exploration. In the next examples, we will analyze the graphs of functions without knowing their formulas.

Example 1

Sketch a graph of the derivative of the function shown in Figure 13.6 on page 228.

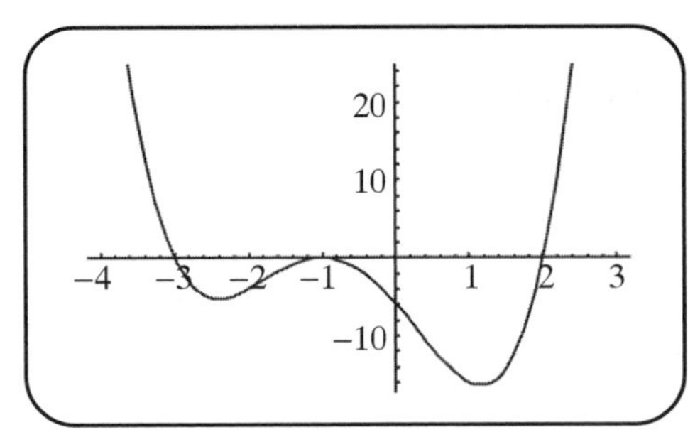

Figure 13.6.

Solution: Start by locating the critical values of the function: The critical points are $x \approx -2.40$, $x \approx -1$, and $x \approx 1.15$. The derivative must equal 0 (and cross the x-axis) at these three points: $f'(-2.40) = 0$, $f'(-1) = 0$, and $f'(1.15) = 0$. Divide the plane into four regions: $x < -2.40$, $-2.40 < x < -1$, $-1 < x < 1.15$, and $1.15 < x$. To sketch the graph of the derivative, start from the left and move to the right. The function is initially decreasing, until $x \approx -2.40$, so $f'(x) < 0$ in the region $x < -2.40$. Then the function begins to increase, until $x \approx -1$, so $f'(x) > 0$ in the region $-2.40 < x < -1$. Between $x \approx -1$ and $x \approx 1.15$, the function is decreasing again, so $f'(x) < 0$ in the region $-1 < x < 1.15$. The rest of the graph shows the function increasing, so $f'(x) > 0$ in the region $1.15 < x$. The sketch of the derivative is shown in Figure 13.7. Your sketch may look a little different, but signs of the derivative must be fairly accurate. The derivative must be above the x-axis when the function is increasing, and it must be below the x-axis when the function is decreasing. The derivative must cross the x-axis at the critical points. The function has local minima at $x \approx -2.40$ and $x \approx 1.15$, so the derivative must change sign from negative to positive around $x \approx -2.40$ and $x \approx 1.15$. The function has a local maximum at $x \approx -1$, so the derivative must change sign from positive to negative around $x \approx -1$.

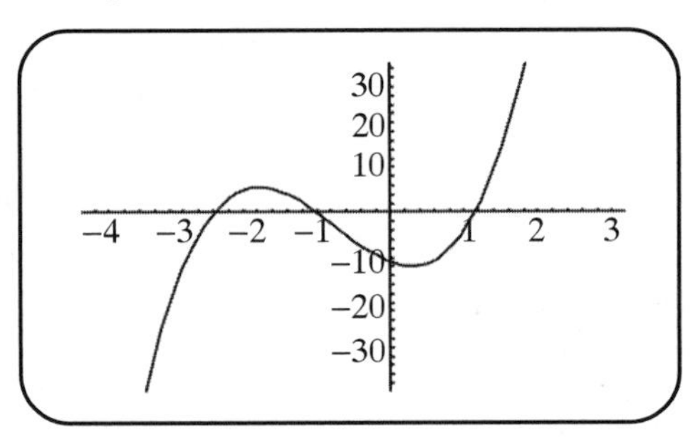

Figure 13.7.

Example 2

Sketch a graph of the derivative of the function shown in Figure 13.8.

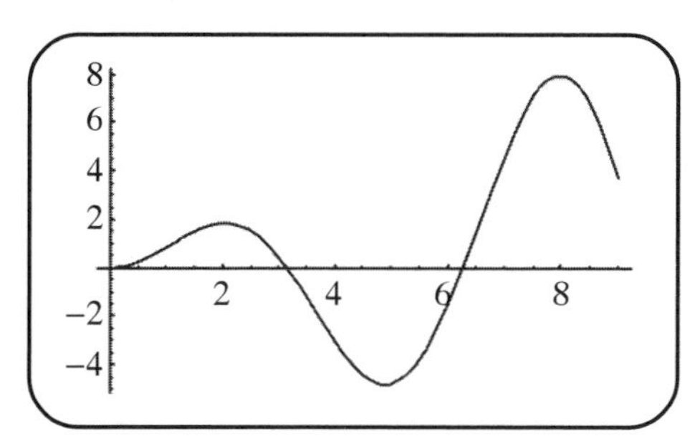

Figure 13.8.

Solution: The function has critical values at $x \approx 2$, $x \approx 5$, and $x \approx 8$. The derivative must equal 0 (and cross the x-axis) at these three points: $f'(2) = 0$, $f'(5) = 0$, and $f'(8) = 0$. Divide the plane into four regions: $0 < x < 2$, $2 < x < 5$, $5 < x < 8$, and $8 < x$. The function is increasing from $x \approx 0$ to $x \approx 2$, so the graph of the derivative will lie above the x-axis on the interval (0, 2). The function is decreasing from $x \approx 2$ to $x \approx 5$, so the graph of the derivative will lie below the x-axis on the interval (2, 5). The function is increasing from $x \approx 5$ to $x \approx 8$, so the graph of the derivative will lie above the axis on the interval (5, 8). Finally, the function is decreasing beyond $x \approx 8$, so the graph of the derivative should be below the x-axis in that region. The sketch of the derivative of the function is shown in Figure 13.9. Again, your sketch may look a little different, and you shouldn't worry so much about the scale. Your emphasis at this point should be on the shape of the graph, the places where the graph touches the x-axis, and the regions where the graph is above or below the x-axis.

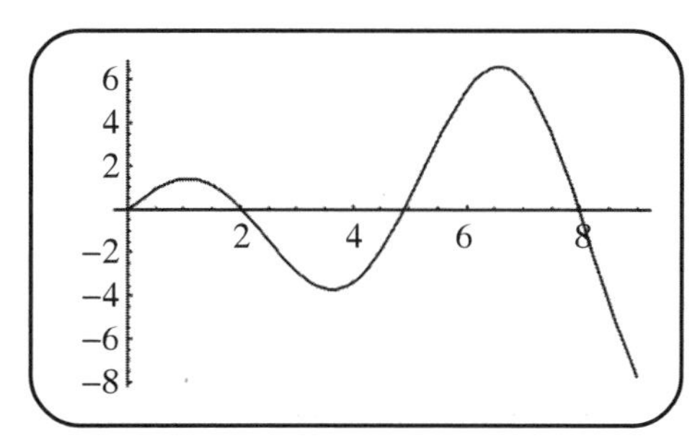

Figure 13.9.

We can also work backwards: Given the graph of the derivative, try to sketch the graph of a possible function that has the given function as its derivative. In other words, given a graph of $f'(x)$, try to graph $f(x)$. In this situation, our answer will not be unique. The important aspects of your sketch will be the shape of the graph and the location of the local extrema. Because you are starting with the graph of $f'(x)$, it will be important to break this graph up into regions where $f'(x)$ is positive (or above the x-axis) and where $f'(x)$ is negative (or below the x-axis). The places where the graph of $f'(x)$ crosses the x-axis (and changes sign) will correspond to local extrema of the graph of $f(x)$. Your sketch of $f(x)$ should be increasing where $f'(x)$ is positive, and it should be decreasing where $f'(x)$ is negative. We will discuss this process in more detail once we discuss the second derivative and concavity.

Lesson 13-3 Review

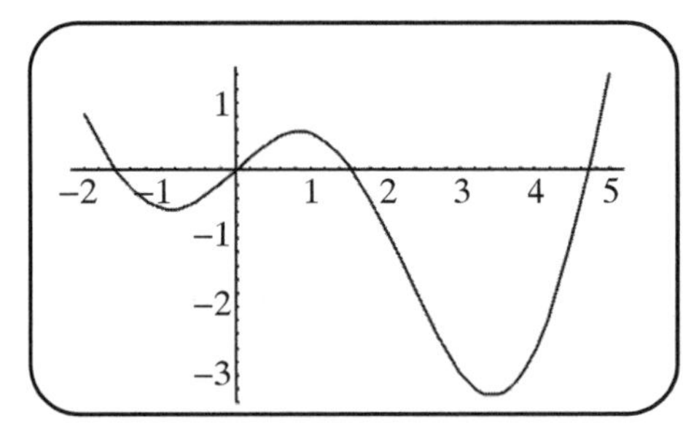

Figure 13.10.

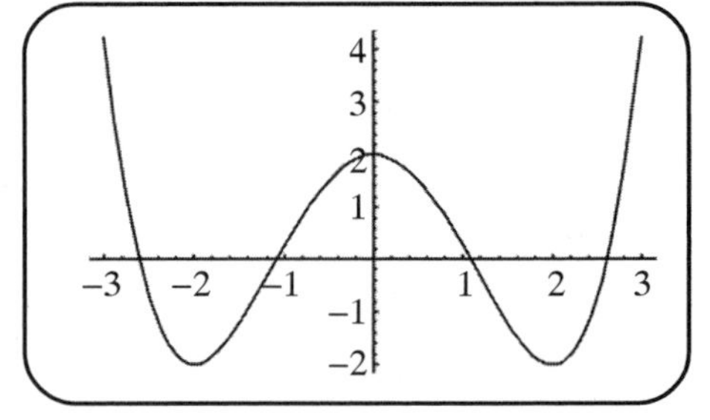

Figure 13.11.

1. Sketch a graph of the derivative of the function shown in Figure 13.10.
2. Sketch a graph of the derivative of the function shown in Figure 13.11.

Answer Key

Lesson 13-1 Review

1. The domain of this function is $[-1, 1]$. Analyze the sign of the derivative:

$$g'(x)=\left(\sqrt{1-x^2}\right)+(x)\left(\tfrac{1}{2}\left(1-x^2\right)^{-1/2}(-2x)\right)$$

$$g'(x)=\left(1-x^2\right)^{-1/2}\left[\left(1-x^2\right)-2x^2\right]$$

$$g'(x)=\left(1-x^2\right)^{-1/2}\left(1-3x^2\right)$$

The sign chart for $g'(x)$ is shown in Figure 13.12:

$g(x)$ is increasing on $\left(-\frac{1}{\sqrt{3}}, \frac{1}{\sqrt{3}}\right)$ and is decreasing on $\left(-1, -\frac{1}{\sqrt{3}}\right) \cup \left(\frac{1}{\sqrt{3}}, 1\right)$.

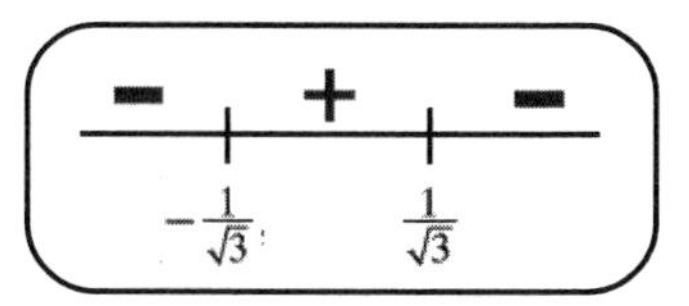

Figure 13.12.

2. $f'(x) = (x+1)e^x$. The sign chart for $f'(x) = (x+1)e^x$ is shown in Figure 13.13: $f(x)$ is increasing on $(-1, \infty)$ and decreasing on $(-\infty, -1)$.

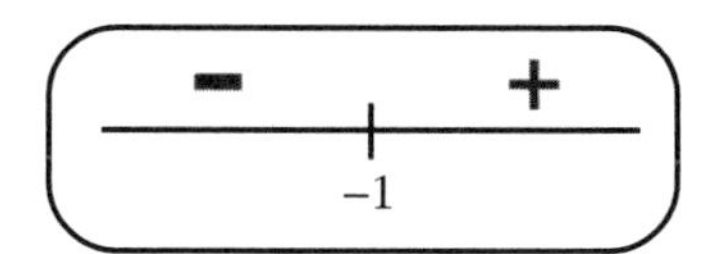

Figure 13.13.

Lesson 13-2 Review

1. $f'(x) = 1 - 2\cos x$ and $f'(x) = 0$ where $\cos x = \frac{1}{2}$. The values of x where $\cos x = \frac{1}{2}$ in $[0, 2\pi]$ are $x = \frac{\pi}{3}$ and $x = \frac{5\pi}{3}$. Evaluate the function at the critical values and at the endpoints: $f(0) = 0$, $f\left(\frac{\pi}{3}\right) = \frac{\pi}{3} - \sqrt{3} \approx -0.685$, $f\left(\frac{5\pi}{3}\right) = \frac{5\pi}{3} + \sqrt{3} \approx 6.97$, and $f(2\pi) = 2\pi \approx 6.28$. The absolute maximum is $\frac{5\pi}{3} + \sqrt{3}$ and occurs at $x = \frac{5\pi}{3}$. The absolute minimum is $\frac{\pi}{3} - \sqrt{3}$ and occurs at $x = \frac{\pi}{3}$. The function has a local maximum at $x = \frac{5\pi}{3}$ and a local minimum at $x = \frac{\pi}{3}$. The graph of $f(x) = x - 2\sin x$ is shown in Figure 13.14.

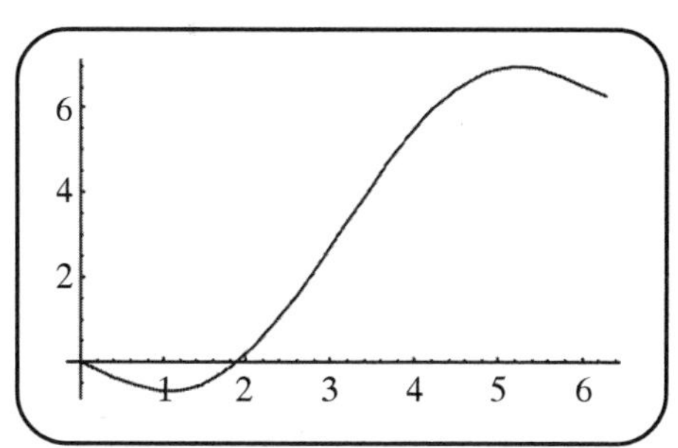

Figure 13.14.

Lesson 13-3 Review

1. The graph the derivative is shown in Figure 13.15.
2. The graph the derivative is shown in Figure 13.16.

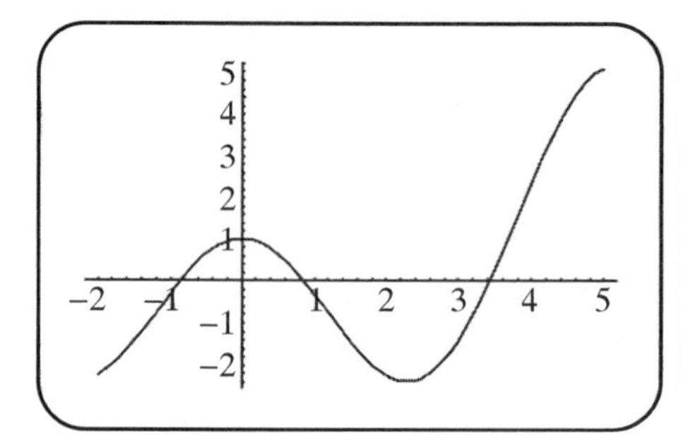

Figure 13.15.

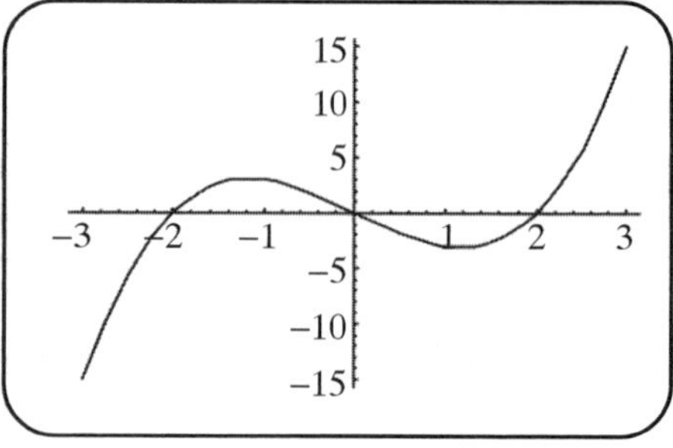

Figure 13.16.

Graphical Analysis Using the Second Derivative

The first derivative is useful in analyzing whether a function is increasing or decreasing. The *way* in which a function is increasing or decreasing is also important. Functions that increase at a *constant* rate are *linear* functions. The shape of a function that is increasing at an increasing rate will be concave up. Whereas the first derivative reflects *how* a function is changing, the second derivative contains information about the shape of a function.

We will use the second derivative to analyze the concavity of the function. The second derivative can also be used to distinguish between local maxima and local minima. This chapter will focus on analyzing functions using the second derivative.

Lesson 14-1: The Second Derivative and Concavity

The second derivative of a function is the derivative of the derivative. Everything we have discussed about the relationship between a function and its derivative can be applied to the relationship between the first derivative of a function and the second derivative of the function. In other words, $f(x)$ is to $f'(x)$ as $f'(x)$ is to $f''(x)$.

We first introduced the idea of concavity when we discussed the graph of a parabola. The shape of a parabola that opens up is concave up, and the shape of a parabola that opens down is concave down. We can actually break a parabola up along its axis of symmetry to obtain two pieces. Both of the pieces that make up a parabola that is concave up will be concave up, and both of the pieces that make up a parabola that is concave down will be concave down. Figure 14.1 on page 234 shows four

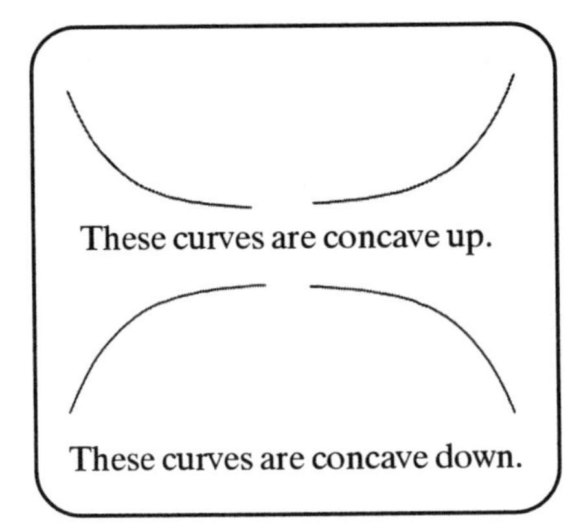

Figure 14.1.

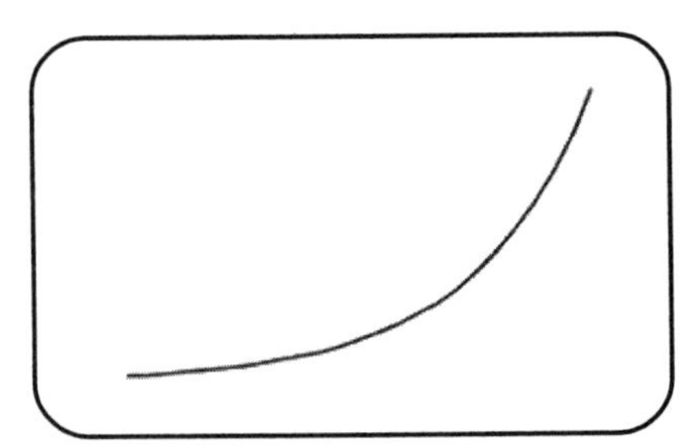

Figure 14.2.

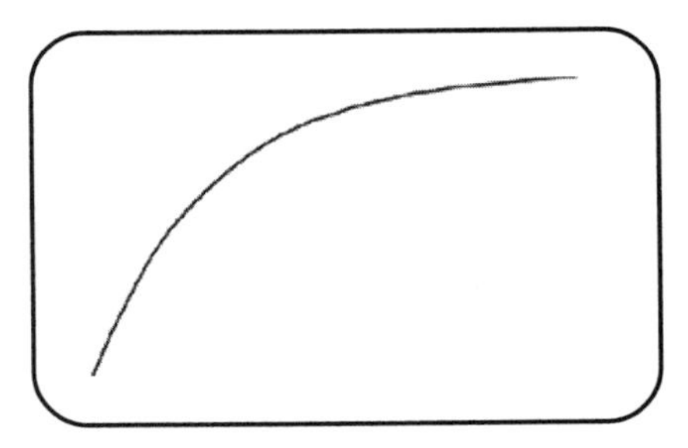

Figure 14.3.

shapes. Two of these shapes are classified as concave up, and two are classified as concave down. We will analyze each one in more detail.

First, consider the function $f(x)$ shown in Figure 14.2. This function is increasing, so $f'(x) > 0$. The slope of the tangent line changes as x changes: As x increases, so does the slope of the tangent line. As x increases, $f'(x)$ increases, so $f'(x)$ is an increasing function, or $f'(x)\uparrow$. The derivative of an increasing function is positive, so the derivative of $f'(x)$ should be positive: $f''(x) > 0$. For the function shown in Figure 14.2, $f(x)\uparrow$ so $f'(x) > 0$, and $f'(x)\uparrow$ so $f''(x) > 0$. For this function, both the derivative and the second derivative are positive. They are working together (in the sense that they both have the same sign), and as a result the function $f(x)$ increases at an increasing rate, or $f(x)$ is increasing concave up.

We will now consider the function $f(x)$ shown in Figure 14.3. This function is also increasing, so $f'(x) > 0$. The slope of the tangent line changes as x changes: As x increases, the slope of the tangent line decreases. As x increases, $f'(x)$ decreases, so $f'(x)$ is a decreasing function, or $f'(x)\downarrow$. The derivative of a decreasing function is negative, so the derivative of $f'(x)$ should be negative: $f''(x) < 0$. For the function shown in Figure 14.3, $f(x)\uparrow$ so $f'(x) > 0$, and $f'(x)\downarrow$ so $f''(x) < 0$. For this function, the derivative is positive, but the second derivative is negative. They are working against each other (in the sense that they have opposite signs), and as a result the function $f(x)$ increases at a decreasing rate, or $f(x)$ is increasing concave down.

Take a look at the function $f(x)$ shown in Figure 14.4. This function is decreasing, so $f'(x) < 0$. The slope of the tangent line changes as x changes: As x increases, the slope of the tangent line is negative and is decreasing in magnitude, which means that the slope of the tangent line is actually increasing. As x increases, $f'(x)$ is negative and increasing, and $f'(x)$ is an increasing function, or $f'(x)\uparrow$. The derivative of an increasing function is positive, so the derivative of $f'(x)$ should be positive: $f''(x) > 0$. For the function shown in Figure 14.4, $f(x)\downarrow$ so $f'(x) < 0$, and $f'(x)\uparrow$ so $f''(x) > 0$. For this function, the derivative is negative, but the second derivative is positive. They are working against each other (in the sense that they have opposite signs), and as a result the function $f(x)$ is decreasing concave up.

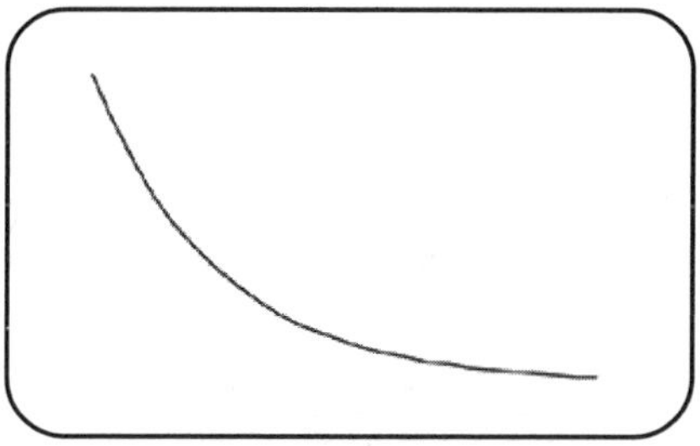

Figure 14.4.

The last graph we will analyze is shown in Figure 14.5. This function is decreasing, so $f'(x) < 0$. The slope of the tangent line changes as x changes: As x increases, the slope of the tangent line is negative and is increasing in magnitude, which means that the slope of the tangent line is actually decreasing. As x increases, $f'(x)$ is negative and decreasing, and $f'(x)$ is a decreasing function: $f'(x)\downarrow$. The derivative of a decreasing function is negative, so the derivative of $f'(x)$ should be negative: $f''(x) < 0$. For the function shown in Figure 14.5, $f(x)\downarrow$ so $f'(x) < 0$, and $f'(x)\downarrow$ so $f''(x) < 0$. For this function, the derivative is negative and the second derivative is also negative. They are working together (in the sense that they have the same sign), and as a result the function $f(x)$ is decreasing concave down.

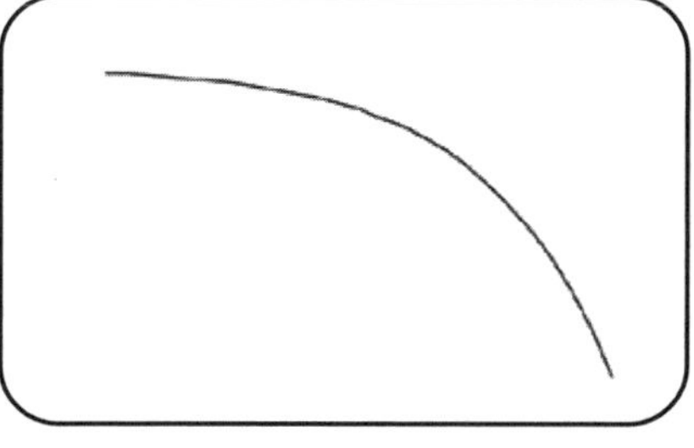

Figure 14.5.

The following table summarizes these observations:

$f(x)$	$f'(x)$	$f''(x)$
$\uparrow \cup$ = Increasing concave up	$+ \uparrow$	$+$
$\uparrow \cap$ = Increasing concave down	$+ \downarrow$	$-$
$\downarrow \cup$ = Decreasing concave up	$- \uparrow$	$+$
$\downarrow \cap$ = Decreasing concave down	$- \downarrow$	$-$

The first derivative can be used to determine the regions where a function is increasing and where it is decreasing. The first derivative can also be used to determine the signs of the second derivative. The second derivative is used to determine where a function is concave up and where it is concave down.

Example 1

Determine the regions where the function $f(x) = xe^x$ is increasing concave up, increasing concave down, decreasing concave up, and decreasing concave down.

Solution: To determine where $f(x) = xe^x$ is increasing or decreasing, we will analyze the signs of $f'(x)$. First, take the derivative using the product rule and simplify:

$$f'(x) = (x)'(e^x) + (x)(e^x)'$$

Use the product rule $\quad f'(x) = (1)(e^x) + (x)(e^x)$

Factor e^x from both terms in the derivative $\quad f'(x) = e^x(1+x)$

Create a sign chart for the derivative.
First, find the critical values: $f'(x) = 0$ when $x = -1$.
Figure 14.6 shows the sign chart for $f'(x)$:
$f'(x) < 0$ on the interval $(-\infty, -1)$, and $f'(x) > 0$ on the interval $(-1, \infty)$.

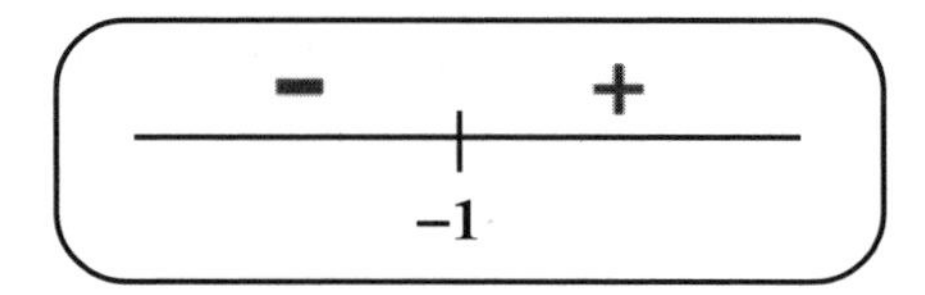

Figure 14.6.

To analyze the concavity of $f(x) = xe^x$, we will need to find the second derivative, or the derivative of $f'(x)$:

$$f''(x) = (e^x)'(x + 1) + (e^x)(x + 1)'$$

Use the product rule $$f''(x) = (e^x)(x + 1) + (e^x)(1)$$

Factor e^x from both terms in the second derivative

$$f''(x) = e^x((x + 1) + 1) = e^x(x + 2)$$

Create a sign chart for the second derivative.
First, find where $f''(x) = 0$: $f''(x) = 0$ at $x = -2$.
Figure 14.7 shows the sign chart for $f''(x)$:
$f''(x) < 0$ on the interval $(-\infty, -2)$ and $f''(x) > 0$ on the interval $(-2, \infty)$.

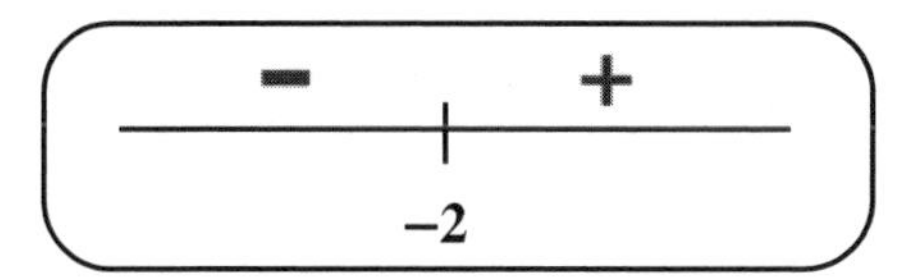

Figure 14.7.

From these two sign charts, we can determine the regions where $f(x)$ is increasing concave up, increasing concave down, decreasing concave up, and decreasing concave down:

Behavior of $f(x)$	$f'(x)$	$f''(x)$	Region
Increasing concave up	+	+	$(-1, \infty)$
Increasing concave down	+	–	None
Decreasing concave up	–	+	$(-2, -1)$
Decreasing concave down	–	–	$(-\infty, -2)$

Lesson 14-1 Review

Determine the regions where the following function is increasing concave up, increasing concave down, decreasing concave up, and decreasing concave down.

1. $f(x) = x \ln x$

Lesson 14-2: Inflection Points

The points where a function changes direction are important. A function has a local maximum when the function makes the transition from increasing to decreasing, and it has a local minimum when it makes the transition from decreasing to increasing. Local extrema are characterized by a change in the sign of the derivative, which means that we can use calculus to find the local extrema of a function. The key step is to find the critical points, or the points where the derivative is equal to 0 or is undefined. Analyzing the signs of the derivative will distinguish between local maxima, local minima, and wild-goose chases.

The points where a function changes concavity, or where the *second* derivative changes sign, are also important points, called points of inflection. A function has an **inflection point** when the shape of the graph changes concavity. An inflection point will occur when the function goes from concave up to concave down, or from concave down to concave up. The role of $f''(x)$ in finding inflection points is analogous to the role that $f'(x)$ played in finding the local extrema. To find the inflection points, find the points in the domain for which $f''(x)$ is either 0 or undefined and examine the behavior of $f''(x)$ around these points. If $f''(x)$ changes sign around any of these points, the function has an inflection point there.

Example 1

Find the inflection points of the function $f(x) = x^3 - 3x^2 + 2x - 1$.

Solution: First, find the second derivative:

$$f'(x) = 3x^2 - 6x + 2$$

$$f''(x) = 6x - 6$$

$$f''(x) = 6(x - 1)$$

Next, create a sign chart for $f''(x)$ and see where it changes sign.
To do this, find the values of x where $f''(x)$ is either 0 or undefined: $f''(x) = 0$ at $x = 1$, and the sign chart for $f''(x)$ is shown in Figure 14.8. The point $x = 1$ is an inflection point of $f(x) = x^3 - 3x^2 + 2x - 1$.

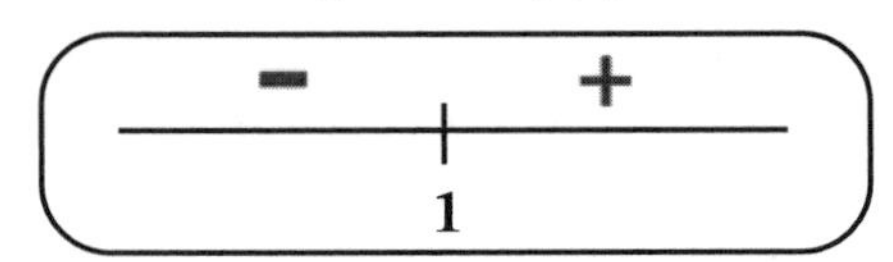

Figure 14.8.

We can use calculus to analyze polynomials of various degrees. The degree of a quadratic function is 2, and the degree of its derivative will be 1: The derivative of the *quadratic* function $f(x) = ax^2 + bx + c$ is the *linear* function $f'(x) = 2ax + b$. All linear functions cross the x-axis once, which means that all quadratic functions will have one point where the derivative is equal to 0 and changes sign. All quadratic functions will have either a local maximum or a local minimum, which is the vertex of the parabola. The formula for finding the x-coordinate of the vertex of a parabola that is taught in algebra, $x = -\frac{b}{2a}$, is a result that is derived using calculus: Set the derivative equal to 0 and solve for x. The second derivative of a quadratic function will be a non-zero constant: $f''(x) = 2a$. A quadratic function does not change concavity and will not have any inflection points.

We can also analyze a cubic function. The degree of a cubic function is 3, and the degree of its derivative will be 2. If $f(x) = ax^3 + bx^2 + cx + d$, then $f'(x) = 3ax^2 + 2bx + c$. A cubic function will have critical values if the corresponding quadratic function $f'(x) = 3ax^2 + 2bx + c$ has real roots. A quadratic function can have either two distinct real roots, one real root of multiplicity 2, or two complex roots. The cubic function will have a local maximum and a local minimum if the quadratic function $f'(x) = 3ax^2 + 2bx + c$ has two distinct real roots. The cubic function will have no local extrema if the quadratic function $f'(x) = 3ax^2 + 2bx + c$ has either one real root of multiplicity 2, or two complex roots. The second derivative of a cubic function is the linear function $f''(x) = 6ax + 2b$. All linear functions have one x-intercept and cross the x-axis once. That means that all cubic functions will have one point where the second derivative is equal to 0 and changes sign. In other words, all cubic functions will have an inflection point, and using calculus we can find the location of the inflection point fairly easily: $x = \frac{-b}{3a}$.

Using calculus, we can see that every polynomial with a positive, even degree will have at least one local maximum or local minimum. Moreover, every polynomial of odd degree greater than 1 will have at least one inflection point. The key to this observation is to remember that every polynomial with odd degree greater than 1 crossed the x-axis at least once. The derivative of a polynomial with even degree will be a polynomial with odd degree, and the second derivative of a polynomial with odd degree greater than 1 will also be a polynomial with odd degree.

Lesson 14-2 Review

Find the inflection points of the following functions:

1. $f(x)=2x^3+5x^2-4x$
2. $h(x)=x+\sin x$

Lesson 14-3: Using the Second Derivative

Another application of the second derivative is in determining whether or not a critical point of a function is a local maximum, local minimum, or neither. This application is called the Second Derivative Test.

The Second Derivative Test

Suppose $f''(x)$ is continuous on an open interval that contains the point $x = c$.
If $f'(c) = 0$ and $f''(c) > 0$, then $f(x)$ has a local minimum at $x = c$.
If $f'(c) = 0$ and $f''(c) < 0$, then $f(x)$ has a local maximum at $x = c$.

There should be little mystery surrounding the Second Derivative Test. If $f'(c) = 0$ and $f''(c) > 0$, then the graph of $f(x)$ is concave up in a small region around $x = c$. The easiest way to visualize a concave up graph is to think of a parabola that opens up. Parabolas that open up have a local minimum. Similarly, if $f'(c) = 0$ and $f''(c) > 0$, then the graph of $f(x)$ is concave down in a small region around $x = c$. The easiest way to visualize a concave down graph is to think of a parabola that opens down. Parabolas that open down have a local maximum.

Example 1

Suppose that $f'(3) = 0$ and $f''(x)=\sqrt{x+1}-\sin(\pi x)-x$. Does the function $f(x)$ have a local maximum, a local minimum, or neither at $x = 3$?

Solution: Use the Second Derivative Test to determine the sign of $f''(x)$:

$f''(3)=\sqrt{3+1}-\sin(3\pi)-3=-1$.

Because $f''(x) < 0$, the point $x = 3$ is a local maximum.

If we construct a new function from an old function, we can use the properties of the old function to help determine the properties of the new function, as we will see in the following example.

Example 2

If $f(x)$ is a positive, increasing, concave up function on an interval I, show that $g(x) = [f(x)]^2$ is also a positive, increasing, concave up function on I.

Solution: Clearly, $g(x) = [f(x)]^2$ is a positive function. In order to show that $g(x) = [f(x)]^2$ is increasing on I, we need to show that $g'(x) > 0$ on I.

Using the chain rule, we see that $g'(x) = 2[f(x)]\,f'(x)$. If $f(x) > 0$ and $f'(x) > 0$ on I, then so is their product, so $g'(x) > 0$ on I, which means that $g(x)$ is increasing on I.

To show that $g(x)$ is concave up on I, we need to show that $g''(x) > 0$ on I. Using the product rule, we see that $g''(x) = 2[f'(x)]\,f'(x) + 2[f(x)]\,f''(x)$. Because $f(x) > 0$, $f'(x) > 0$, and $f''(x) > 0$ on I (we are given that $f(x)$ is a positive, increasing, concave up function), $g''(x) > 0$ on I.

Therefore, $g(x)$ is also a positive, increasing, concave up function on I.

Throughout this book we have practiced finding tangent line equations. The point of finding tangent line equations is that we can use the tangent line to approximate the function in a small region around the intersection of the function and its tangent line. When we approximate a function using its tangent line, it is important to know whether our approximation is an underestimate (meaning that our estimate is smaller than the function) or an overestimate (meaning that our estimate is larger than the function). The second derivative will help us determine whether our approximation is an overestimate or an underestimate.

If the graph of a function is concave up, then the tangent line will lie below the function, and any estimate we give using the tangent line will be too small. If the graph of a function is concave down, the tangent line will lie above the function, and any estimate we give using the tangent line will be too large.

Example 3

Find the equation of the line tangent to the graph of the function $e^y + \sin y = e^{2x} + x$ at the origin, and use the tangent line to estimate the function at $x = 0.1$. Is your approximation for y an overestimate or an underestimate?

Solution: We need to differentiate this function implicitly:

$e^y + \sin y = e^{2x} + x$

Differentiate each term using the chain rule where appropriate

$$e^y y' + \cos y y' = 2e^{2x} + 1$$

Factor y' from both terms on the left $\quad y'(e^y + \cos y) = 2e^{2x} + 1$

Divide by $(e^y + \cos y)$ $\quad y' = \dfrac{2e^{2x} + 1}{(e^y + \cos y)}$

Now, evaluate y' when $x = 0$ and $y = 0$ $\quad y' = \dfrac{2e^0 + 1}{(e^0 + \cos 0)} = \dfrac{3}{2}$

Next, use the point-slope formula to find the equation of the tangent line

$$y - 0 = \tfrac{3}{2}(x - 0)$$
$$y = \tfrac{3}{2}x$$

Use the tangent line to estimate the value of the function when $x = 0.1$: $y = 0.15$.

To determine whether this is an overestimate or an underestimate, find the second derivative by differentiating $y' = \dfrac{2e^{2x} + 1}{(e^y + \cos y)}$ implicitly:

$$y' = \frac{2e^{2x} + 1}{(e^y + \cos y)}$$

Use the quotient rule

$$y'' = \frac{(2e^{2x} + 1)'(e^y + \cos y) - (2e^{2x} + 1)(e^y + \cos y)'}{(e^y + \cos y)^2}$$

Differentiate each term, and use the chain rule where necessary

$$y'' = \frac{(4e^{2x})(e^y + \cos y) - (2e^{2x} + 1)(e^y y' - \sin y y')}{(e^y + \cos y)^2}$$

Evaluate y'' when $x = 0, y = 0$ and $y' = 1$

$$y'' = \frac{(4e^0)(e^0 + \cos 0) - (2e^0 + 1)(e^0 \cdot \frac{3}{2} - (\sin 0) \cdot \frac{3}{2})}{(e^0 + \cos 0)^2} = \frac{4 - 3}{2^2} = \frac{1}{4}$$

Because $y'' > 0$, when $x = 0$, $y = 0$ and $y' = 1$, the graph of the function is concave up, so the tangent line lies below the function and we have an underestimate.

One way to find the exact value of the function $e^y + \sin y = e^{2x} + x$ at $x = 0.1$ is to graph the functions $f(x) = e^x + \sin x$ and $g(x) = e^{0.2} + 0.1$ on the same axes. The x-coordinate of the point of intersection will be the exact value of $e^y + \sin y = e^{2x} + x$ at $x = 0.1$. The graphs of these two functions are shown in Figure 14.9. The point of intersection is $x \approx 0.155$, which is close to, and larger than, our estimate, which agrees with our answer.

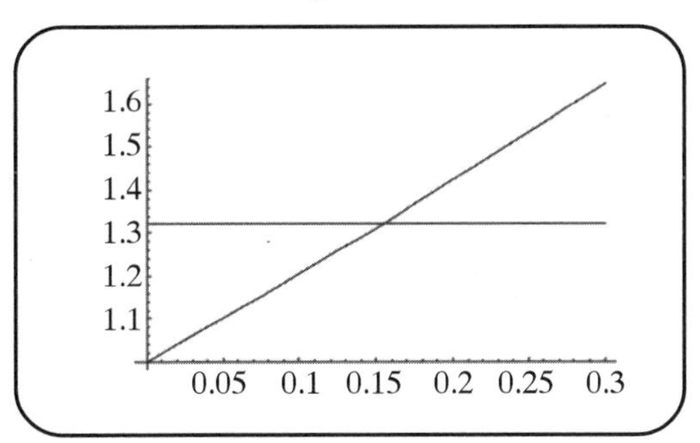

Figure 14.9.

Lesson 14-3 Review

1. Suppose that $f(3) = -2$, $f'(3) = 0$, and $f''(x) = \sqrt{x+6} - \cos(\pi x) - x^2$. Does $f(x)$ have a local maximum, a local minimum, or neither at $x = 3$?

2. Find the equation of the line tangent to the graph of the function $\cos x + x = e^y + y$ at the origin, and use the tangent line to estimate the function at $x = 0.1$. Is your approximation for y an overestimate or an underestimate?

Lesson 14-4: Relating the Graphs of $f(x)$, $f'(x)$, and $f''(x)$

Now that we have some insight into the information contained in the first and second derivatives, we can put these ideas together and relate the graphs of $f(x)$, $f'(x)$ and $f''(x)$. One of our goals is to be able to understand the behavior of $f(x)$ if we are given graphs of either $f'(x)$ or $f''(x)$. We will also want to understand the graph of $f''(x)$ if we are given the graph of $f'(x)$.

The graph of the derivative of a function, $f'(x)$, contains information about the graph of the function $f(x)$. Wherever the graph of $f'(x)$ lies above the x-axis, $f(x)$ is increasing, and wherever the graph of $f'(x)$ lies below the x-axis, $f(x)$ is decreasing. The places where the graph of $f'(x)$ crosses the x-axis will correspond to local extrema of $f(x)$. The regions where $f'(x)$ is increasing indicate that the graph of $f(x)$ is concave up, and the regions where $f'(x)$ is decreasing indicate that the graph of $f(x)$ is concave down. Remember that if $f'(x)$ is increasing, then $f''(x) > 0$, which means that $f(x)$ is concave up. If $f'(x)$ is decreasing, then $f''(x) < 0$, which means that $f(x)$ is concave down. The local extrema of the graph of $f'(x)$ will correspond to the inflection points of $f(x)$.

The graph of the second derivative of a function, $f''(x)$, also contains information about the graph of $f(x)$, but not as much as the graph of $f'(x)$. We can use the graph of $f''(x)$ to extract information about the concavity of the graph of $f(x)$, and we can determine the points of inflection.

Example 1

Given the graph of $f'(x)$ as shown in Figure 14.10, indicate the regions where $f(x)$ is increasing.

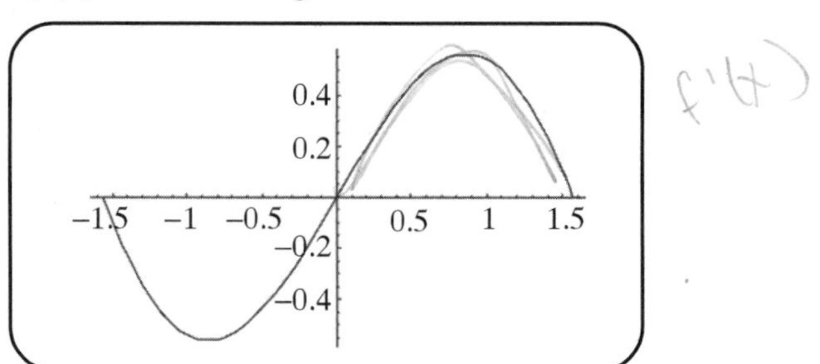

Figure 14.10.

Solution: If $f(x)$ is increasing, then $f'(x) > 0$, or the graph of $f'(x)$ will be above the x-axis. The region where $f(x)$ is increasing is shown in Figure 14.11.

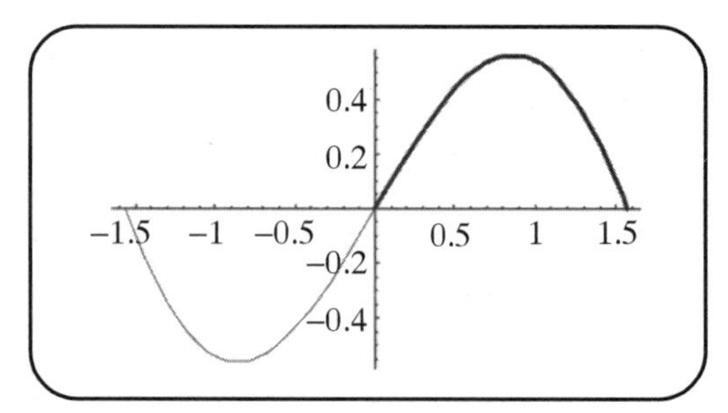

Figure 14.11.

Example 2

Given the graph of $f'(x)$ as shown in Figure 14.12, indicate the regions where $f(x)$ is concave down.

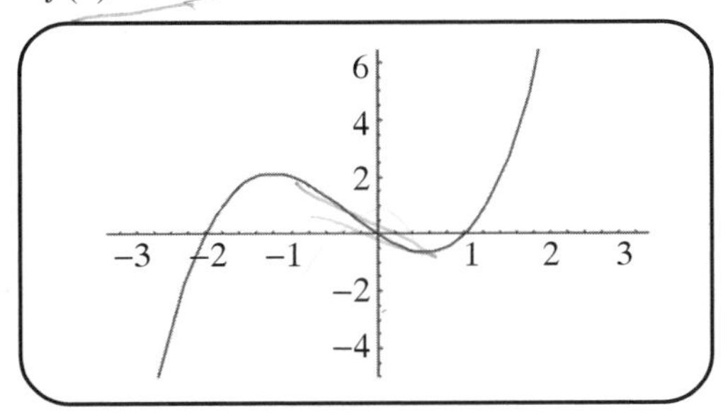

Figure 14.12.

Solution: If $f(x)$ is concave down, then $f''(x) < 0$, which means that $f'(x)$ is decreasing; $f'(x)$ is decreasing on $(-1.2, 0.6)$. The region where $f(x)$ is concave down is shown in Figure 14.13.

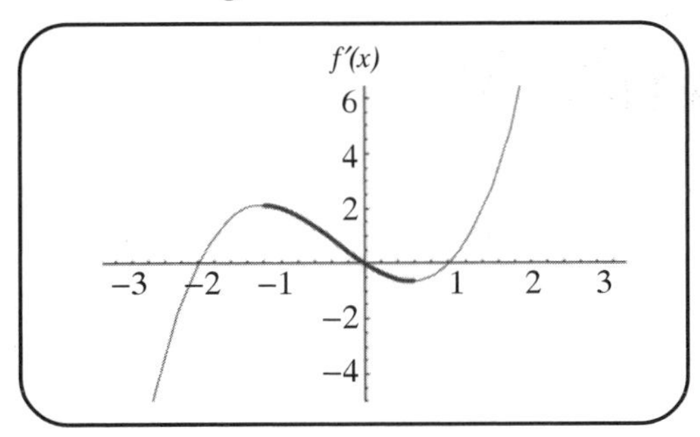

Figure 14.13.

Example 3

Given the graph of $f'(x)$ as shown in Figure 14.14, indicate the location of any local maxima of $f(x)$.

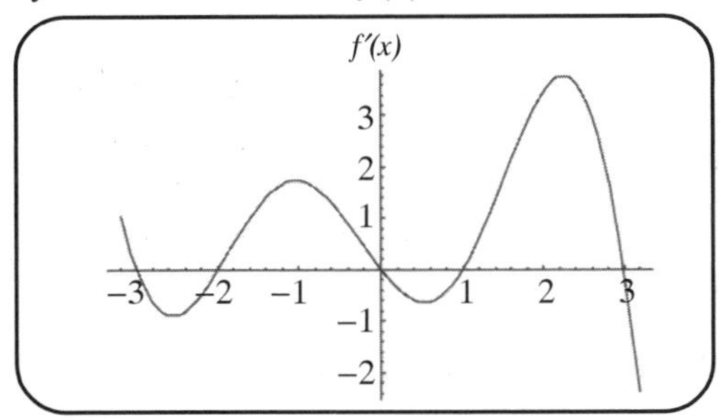

Figure 14.14.

Solution: The location of the local maxima of $f(x)$ will occur where the graph of $f'(x)$ crosses the x-axis and changes sign from *positive* to *negative*, which are the points $x = -3$, $x = 0$, and $x = 3$. The location of the local maxima of $f(x)$ is shown in Figure 14.15.

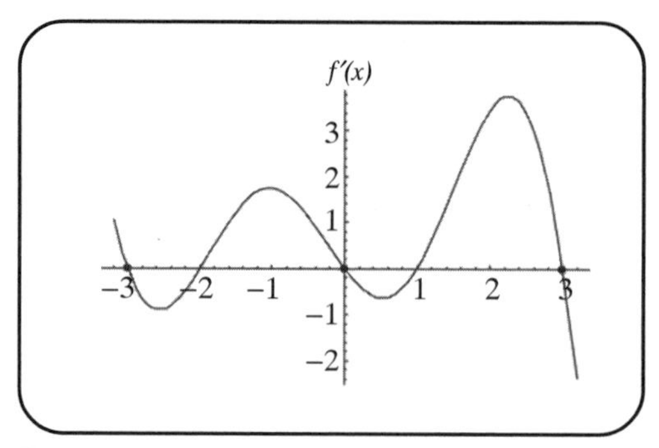

Figure 14.15.

Example 4

Given the graph of $f'(x)$ as shown in Figure 14.16, indicate the location of the inflection points of $f(x)$.

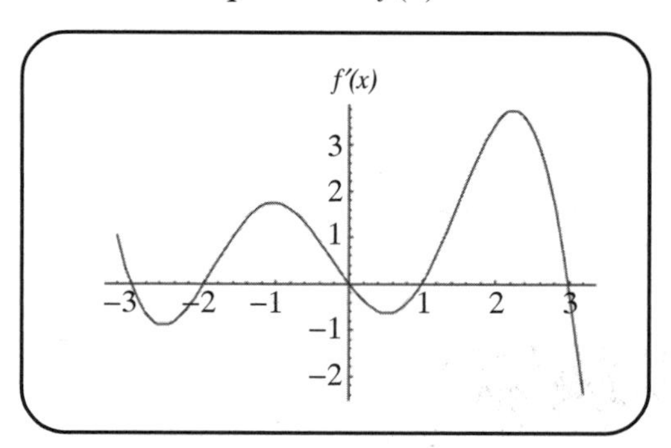

Figure 14.16.

Solution: $f(x)$ will have inflection points where the graph of $f'(x)$ has local maxima or local minima. The local extrema of $f'(x)$ are the locations where $f''(x) = 0$ and changes sign, which means that they are the inflection points of $f(x)$. The location of the inflection points of $f(x)$ is shown in Figure 14.17 on page 247.

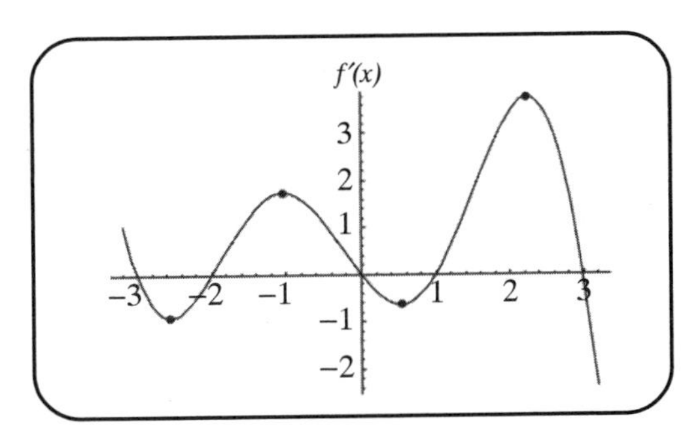

Figure 14.17.

Example 5

Given the graph of $f''(x)$ as shown in Figure 14.18, indicate the regions where $f(x)$ is concave up.

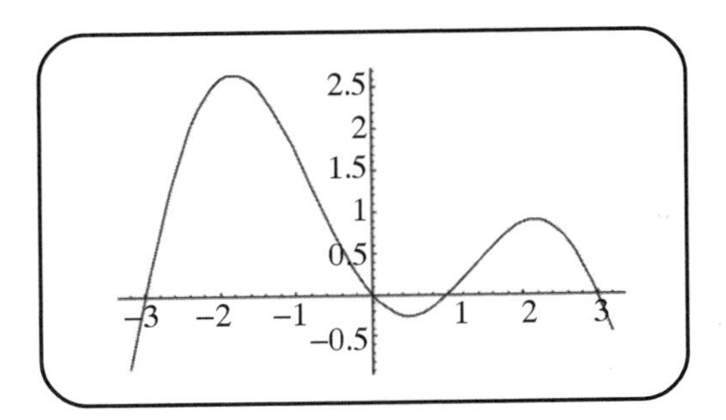

Figure 14.18.

Solution: The regions where $f(x)$ is concave up are the regions where $f''(x) > 0$, or where the graph of $f''(x)$ is above the x-axis. The regions where $f(x)$ is concave up are shown in Figure 14.19.

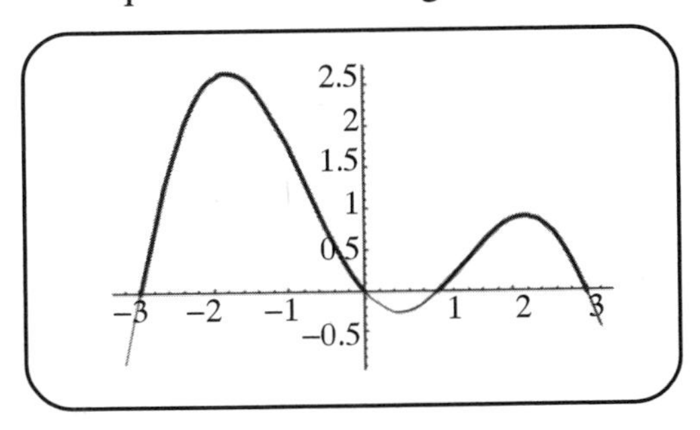

Figure 14.19.

Lesson 14-4 Review

1. Given the graph of $f(x)$ as shown in Figure 14.20, indicate the regions where $f'(x)$ is increasing.

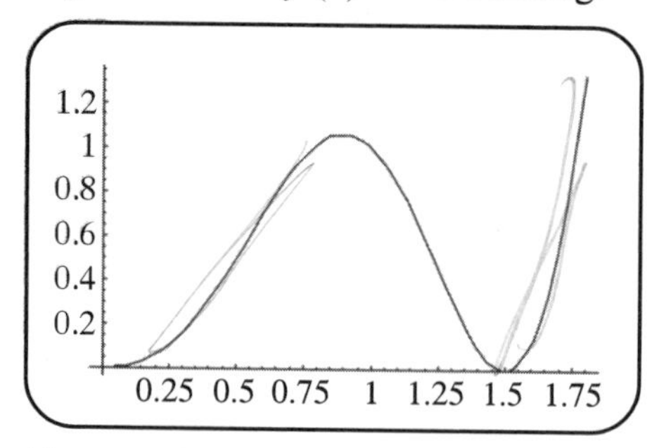

Figure 14.20.

2. Given the graph of $f'(x)$ as shown in Figure 14.21, indicate the regions where $f''(x) > 0$.

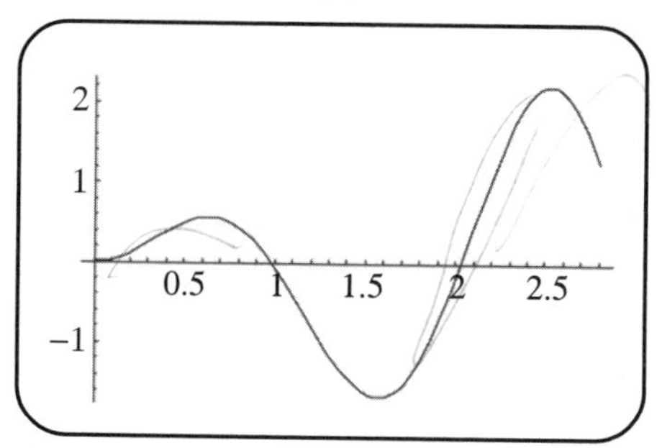

Figure 14.21.

3. Given the graph of $f''(x)$ as shown in Figure 14.22, indicate the location of any inflection points of $f(x)$.

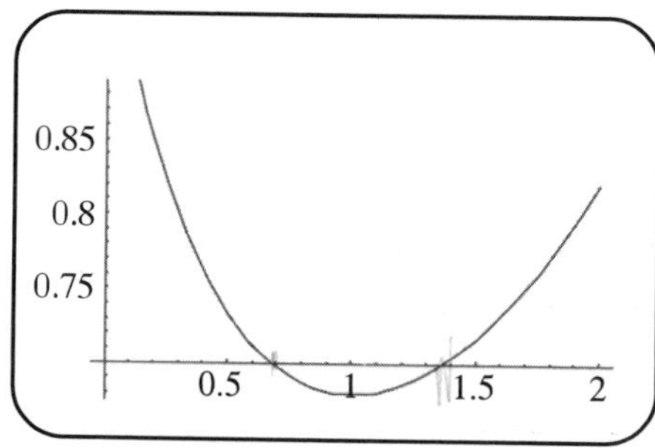

Figure 14.22.

Answer Key

Lesson 14-1 Review

1. $f'(x) = \ln x + 1 : f'(x) > 0$ on $\left(\frac{1}{e}, \infty\right)$, $f'(x) < 0$ on $\left(0, \frac{1}{e}\right)$

   $f''(x) = \frac{1}{x} : f''(x) > 0$ on $(0, \infty)$

   $f(x)$ is increasing concave up on $\left(\frac{1}{e}, \infty\right)$, decreasing concave up on $\left(0, \frac{1}{e}\right)$.
   It is never concave down.

Lesson 14-2 Review

1. $f'(x) = 6x^2 + 10x - 4$, $f''(x) = 12x + 10$;

   $f''(x) = 0$ at $x = -\frac{5}{6}$, $f''(x) < 0$ on $\left(-\infty, -\frac{5}{6}\right)$ and $f''(x) > 0$ on $\left(-\frac{5}{6}, \infty\right)$,

   so $f(x)$ has an inflection point at $x = -\frac{5}{6}$.

2. $h'(x) = 1 + \cos x$, $h''(x) = -\sin x$. The inflection points will occur when $h''(x) = 0$ and changes sign, so the inflection points are located at $(n\pi, n\pi)$.

Lesson 14-3 Review

1. $f''(3) = -5$, so by the Second Derivative Test, $f(x)$ has a local maximum at $x = 3$.

2. $y' = \frac{1-\sin x}{e^y + 1}$, and when $x = 0$ and $y = 0$, $y' = \frac{1}{2}$.

   The equation of the tangent line is $y = \frac{1}{2}x$, so when $x = 0.1$, $y \approx 0.05$.

   $y'' = \frac{(-\cos x)(e^y + 1) - (1 - \sin x)(e^y y')}{(e^y + 1)^2}$, so when $x = 0$, $y = 0$, and $y' = \frac{1}{2}$, $y'' = -\frac{5}{8} < 0$, so

   the graph is concave down at the origin and our approximation is an overestimate.

Lesson 14-4 Review

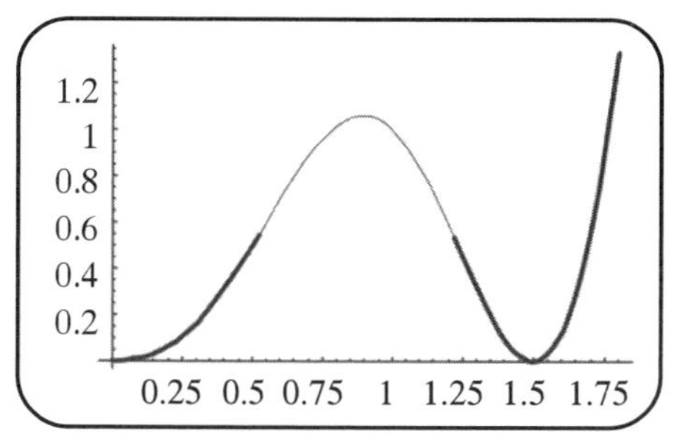

Figure 14.23.

1. The regions where $f'(x)$ is increasing are shown in Figure 14.23

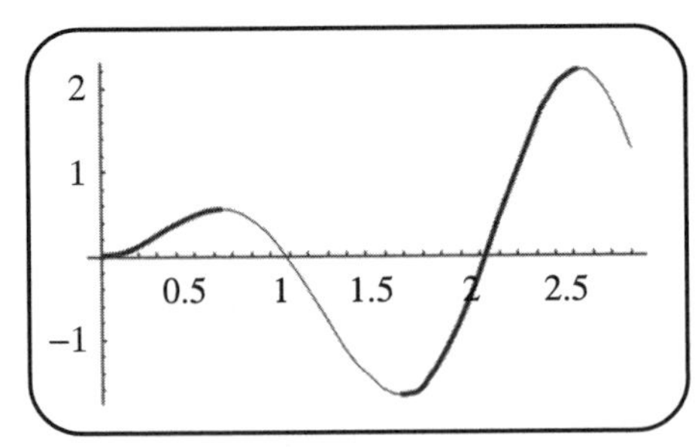

Figure 14.24.

2. The regions where $f''(x) > 0$ are shown in Figure 14.24
3. The location of the inflection points of $f(x)$ is shown in Figure 14.25.

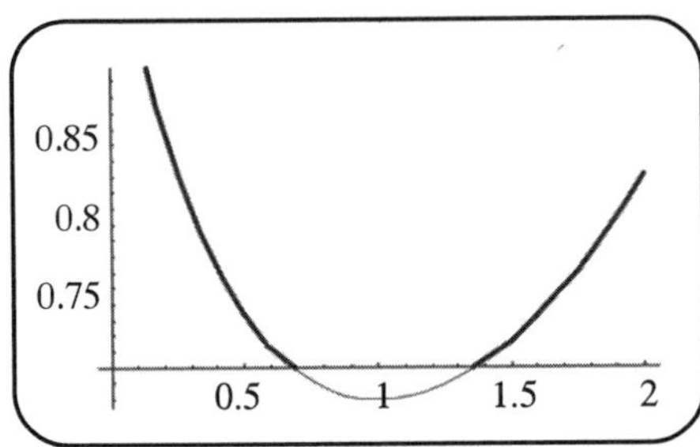

Figure 14.25.

Integration

I mentioned earlier that calculus was the study of limits. The derivative is a limit of difference quotients, and we have seen many applications of the derivative. We can use the derivative to find equations of tangent lines, analyze the graph of a function, solve optimization problems, and quantify related rates. The difference quotient measures the change in a function relative to the change in its independent variable. The branch of calculus that is focused on the derivative is called *differential calculus*. Differential calculus is not the only type of calculus. There is another branch of calculus, called *integral calculus*, that we will now explore.

Lesson 15-1: Anti-Derivatives

Anti-differentiation is the reverse of differentiation. With differentiation, you are given a function and you have to take its derivative, either by applying the product, quotient, or chain rule. With **anti-differentiation**, or **integration**, you are given the derivative and you must find the original function. The nice thing about anti-differentiation is that you can always check your work. Differentiating the anti-derivative should bring you back to square one. In this lesson we will limit ourselves to finding anti-derivatives of elementary functions.

Suppose the derivative of a function is given by the formula $f'(x) = 2x$. We will denote the anti-derivative, or integral, of $2x$ as $\int 2x dx$. The symbol $\int$ is called an **integral**, and $\int 2x dx$ is called an **indefinite integral**. The function within the integral, which in this case is $2x$, is called the **integrand**. An indefinite integral and an anti-derivative refer to the same thing. To find the anti-derivative of $2x$, we need to find a function whose *derivative* is $2x$.

One such function is $f(x) = x^2$. We can check: Is the derivative of $f(x) = x^2$ equal to $2x$? Yes, it is. However, someone else may speak up and say that $f(x) = x^2 + 10$ is also a function whose derivative is equal to $2x$. They would also be correct. In fact, if c is any constant, then the derivate of $f(x) = x^2 + c$ will be equal to $2x$. The anti-derivative, or integral, of a function is not unique; it will be a class of functions that differ by a constant. In general, we write $\int 2x dx = x^2 + c$, where c is called the **integration constant.**

There are a few elementary functions that are fairly straightforward to integrate. If you think about integration in terms of undoing differentiation (so that the integral is the *inverse* of differentiation, in some sense), then thinking about the differentiation process should help with understanding the integration process. To take the derivative of a power function, the exponent comes down as a product, and then the exponent of the power function is reduced by 1. The inverse of this process would involve reversing these steps in reverse order: First increase the exponent of the power function by 1, and then divide by this new power. Because the answer is not unique, we need to include the integration constant:

$$\int x^n dx = \frac{x^{n+1}}{n+1} + c$$

There is only one problem with this formula, and that occurs when $n = -1$. When $n = -1$, the expression $\frac{x^{n+1}}{n+1} + c$ makes no sense, as the denominator is 0. In order to integrate the function $x^{-1} = \frac{1}{x}$, we need to think of a function whose derivative is $\frac{1}{x}$. Remember that the derivative of $f(x) = \ln x$ is $f'(x) = \frac{1}{x}$. In general, to integrate a power function, we have the following rule:

$$\int x^n dx = \begin{cases} \frac{x^{n+1}}{n+1} + c & n \neq -1 \\ \ln|x| & n = -1 \end{cases}$$

The reason for the absolute value symbols in the logarithm has to do with the domains of the functions $\frac{1}{x}$ and $\ln x$: The domain of $\frac{1}{x}$ is the set of all real numbers such that $x \neq 0$, and the domain of $\ln x$ is the set of all positive real numbers. Including the absolute value of x in the argument of the logarithm expands the domain of $\ln x$ to match that of $\frac{1}{x}$: the domain of $\ln |x|$ is the set of all real numbers such that $x \neq 0$.

There are other elementary functions that we can integrate. Because the derivative of $f(x) = e^x$ is e^x, we also know that: $\int e^x dx = e^x + c$. We can also integrate trigonometric and hyperbolic functions from the formulas for their derivatives:

Formula for the Derivative	Formula for the Integral
$D_x(\sin x) = \cos x$	$\int \cos x dx = \sin x + c$
$D_x(\cos x) = -\sin x$	$\int \sin x dx = -\cos x + c$
$D_x(\tan x) = \sec^2 x$	$\int \sec^2 x dx = \tan x + c$
$D_x(\sec x) = \sec x \tan x$	$\int \sec x \tan x dx = \sec x + c$
$D_x(\csc x) = -\csc x \cot x$	$\int \csc x \cot x dx = -\csc x + c$
$D_x(\cot x) = -\csc^2 x$	$\int \csc^2 x dx = -\cot x + c$
$D_x(\sinh x) = \cosh x$	$\int \cosh x dx = \sinh x + c$
$D_x(\cosh x) = \sinh x$	$\int \sinh x dx = \cosh x + c$
$D_x(\tan^{-1} x) = \frac{1}{1+x^2}$	$\int \frac{1}{1+x^2} dx = \tan^{-1} + c$

In general, if the formula for the derivative is given by $D_x(f(x)) = f'(x)$, then the formula for the integral is $\int f'(x)dx = f(x) + c$.

If the derivative of $f(x) = \sin(x^2 + 1)$ is $f'(x) = 2x\cos(x^2 + 1)$, then $\int 2x \sin(x^2 + 1)dx = \cos(x^2 + 1) + c$. The key to successful integration is really knowing your derivatives!

Because integration and differentiation are so closely related, *some* of the properties of derivatives also hold for integrals. In particular, if c is a constant, then: $D_x(c \cdot f(x)) = c \cdot D_x(f(x))$.

From this, we see that: $\int(c\cdot f(x))dx = c\cdot\int f(x)dx$.

Also, because: $D_x(f(x)\pm g(x)) = D_x(f(x)) + D_x(g(x))$,

we have: $\int(f(x)\pm g(x))dx = \int f(x)dx + \int g(x)dx$.

With these two observations, we can integrate the sum of power, trigonometric, exponential, or hyperbolic functions.

Example 1

Evaluate the following indefinite integrals:

a. $\int(x^3+\sin x)dx$

b. $\int(3x+2\sqrt{x})dx$

c. $\int(4e^x+\cos x)dx$

d. $\int\left(\frac{3}{x^2}+\frac{2}{x}\right)dx$

Solution: Use the rules for integrating elementary functions, and let the constants just hang out. Be sure to check your answers by differentiating them, and remember the integration constant:

a. $\int(x^3+\sin x)dx = \frac{x^4}{4} - \cos x + c$

b. $\int(3x+2\sqrt{x})dx = \int(3x+2x^{1/2})dx$

$$= 3\left(\frac{x^2}{2}\right) + 2\left(\frac{x^{3/2}}{3/2}\right) + c = \frac{3}{2}x^2 + \frac{4}{3}x^{3/2} + c$$

c. $\int(4e^x+\cos x)dx = 4e^x + \sin x + c$

d. $\int\left(\frac{3}{x^2}+\frac{2}{x}\right)dx = \int(3x^{-2}+2x^{-1})dx$

$$= 3\left(\frac{x^{-1}}{-1}\right) + 2\ln|x| + c = \frac{-3}{x} + 2\ln|x| + c$$

Lesson 15-1 Review

Evaluate the following indefinite integrals:

1. $\int(3x^4+\sinh x)dx$
2. $\int\left(\frac{3}{\sqrt{x}}+2\sec^2 x\right)dx$
3. $\int(2e^x+\sec x\tan x)dx$
4. $\int\left(\frac{3}{x^5}-\frac{4}{x}\right)dx$

Lesson 15-2: Substitution

The anti-differentiation formulas given in the previous lesson will *not* help us evaluate integrals of the form $\int \frac{1}{x+2}dx$ or $\int 2x\sqrt{x^2+1}dx$. To integrate these types of functions, we need to simplify the integrand so that it matches one of the formulas in our table. To apply the technique correctly, every instance of the old variable must be removed before any integration can take place. In the integral $\int f(x)dx$, the variable makes two appearances: One appearance is in the formula for the integrand, $f(x)$, and the other is in the term dx. We must eliminate the variable in *both* of its locations. Once we have integrated the function, we will need to put everything back in its original form by substituting back into the original variable.

Example 1

Evaluate $\int \frac{1}{x+2}dx$.

Solution: We *can* integrate $\int \frac{1}{x}dx$: $\int \frac{1}{x}dx = \ln|x| + c$ (keeping in mind that the variable x is a "dummy" variable in that it does not matter which letter of the alphabet we use to represent the variable, as long as we are consistent throughout), so if our integrand could be turned into a function of this form, we could integrate it. If we let $u = x + 2$, then the function in the integrand will become $\frac{1}{u}$. The equation $u = x + 2$ is called the **substitution equation**. We also must substitute in for dx. To do this, differentiate both sides of the substitution equation using differential notation: $du = dx$. Make the substitutions, integrate, and then substitute back to the original variable:

$$\int \frac{1}{x+2}dx = \int \frac{1}{u}du = \ln|u| + c = \ln|x+2| + c.$$

You can always differentiate your answer (using the chain rule) to see if your answer is correct.

Integration by substitution is the integration equivalent of the chain rule for derivatives. Any time you use the substitution technique for integration, checking your answer by differentiation should involve using the chain rule.

Example 2

Evaluate $\int 2x\sqrt{x^2+1}dx$.

Solution: The problem with this integrand is that it is not a power function because of the quantity ($x^2 + 1$) under the radical. The quantity ($x^2 + 1$) is what ruins our ability to integrate the function, so that is what we will want to eliminate through substitution. Let $u = x^2 + 1$. We also need to differentiate the substitution equation: $du = 2xdx$. To help you keep track of the substitutions, it may help to rearrange the integrand:

$$\int 2x\sqrt{x^2+1}dx = \int \sqrt{x^2+1}(2xdx)$$

Make the substitutions, integrate, and then clean things up:

$$\int 2x\sqrt{x^2+1}dx = \int \sqrt{x^2+1}(2xdx) = \int \sqrt{u}du = \int u^{1/2}du = \frac{u^{3/2}}{3/2} + c$$

$$= \tfrac{2}{3}\left(x^2+1\right)^{3/2} + c$$

We can verify our answer by differentiating the function $\frac{2}{3}\left(x^2+1\right)^{3/2} + c$.

In general, integration by substitution works with integrals of the form $\int F(g(x))g'(x)dx$. The substitution $u = g(x)$ and $du = g'(x)dx$ turns this integral into one of the form $\int F(u)du$, and if $F(u)$ is a function on our list of anti-derivatives, we can then evaluate the integral.

Example 3

Evaluate $\int \frac{(\ln x)^2}{x}dx$.

Solution: We do not have any formulas for when the integrand is a logarithmic function, so that is a big clue as to how to proceed. Also, the x in the denominator can be thought of as $\frac{1}{x}$, which happens to be the derivative of the logarithmic function. Use the substitution $u = \ln x$, and $du = \frac{1}{x}dx$: $\int \frac{(\ln x)^2}{x}dx = \int (\ln x)^2\left(\frac{1}{x}dx\right) = \int u^2 du = \frac{u^3}{3} + c = \frac{1}{3}\left(\ln|x|\right)^3 + c$

Example 4

Evaluate $\int \frac{2x+3}{x^2+3x+1}dx$.

Solution: In this problem, the function in the numerator is equal to the *derivative* of the function in the denominator, so we should substitute in for the function in the denominator. The numerator will vanish along with the dx: $u = x^2 + 3x + 1$, $du = (2x + 3)dx$. Substituting into the integral, we have:

$$\int \frac{2x+3}{x^2+3x+1}dx = \int \left(\frac{1}{x^2+3x+1}\right)(2x+3)dx = \int \frac{1}{u}du = \ln|u| + c = \ln|x^2+3x+1| + c$$

The hardest part about integration by substitution is determining the substitution to make. I have written some integrals and a substitution that would simplify the problem to give a function that can be integrated directly. As you look at the problems, try to come up with your own substitution and see if it matches mine. You do not have to take my substitution advice; there are often several different approaches that would work. Keep in mind that the goal is to be able to evaluate the integral, not just to match my substitution!

Integral	Substitution
$\int \frac{e^x}{e^x+1}dx$	$u = e^x + 1$, $du = e^x dx$
$\int \frac{\tan^{-1}x}{x^2+1}dx$	$u = \tan^{-1}x$, $du = \frac{1}{x^2+1}dx$
$\int \frac{x}{1+x^4}dx$	$u = x^2$, $du = 2xdx$
$\int \frac{\sin x}{\cos^2 x}dx$	$u = \cos x$, $du = -\sin xdx$
$\int \frac{\cos\sqrt{x}}{\sqrt{x}}dx$	$u = \sqrt{x}$, $du = \frac{1}{2\sqrt{x}}dx$

Lesson 15-2 Review

Evaluate the following integrals:

1. $\int \sin(2x+3)dx$

2. $\int \frac{\left(1+\sqrt{x}\right)^9}{\sqrt{x}}dx$

3. $\int x\sqrt{x-1}dx$

Lesson 15-3: Integration by Parts

The chain rule was the basis for the substitution technique for integration. The product rule for differentiation is the motivation for the method of integration by parts. The formula for integration by parts is:

$$\int u(x)v'(x)dx = u(x)v(x) - \int u'(x)v(x)dx$$

This technique is commonly used when the integrand involves a combination of a polynomial function and either an exponential, trigonometric, logarithmic, or hyperbolic function. In the integration by parts process, the integrand will be treated as the product of two functions, $u(x)$ and $v'(x)$. In this method, $u(x)$ is differentiated and $v'(x)$ is integrated. The hope is that the integral on the right is easier to evaluate than the integral on the left. The difficult part about this method is breaking the integrand into the product of two functions, one of which will be differentiated, the other integrated. Differentiating $u(x)$ should be no problem, but you *must* be able to integrate the function that is defined to be $v'(x)$.

The property of polynomials that is exploited in this method is that differentiating a polynomial reduces the degree of the polynomial by 1. Eventually, all polynomials differentiate to a constant. If the integrand involves the product of a polynomial with another type of function, the typical approach is to let $u(x)$ equal the polynomial and $v'(x)$ equal the other part of the integrand.

Example 1

Evaluate $\int x\cos x dx$.

Solution: This integrand involves a product of a polynomial and a trigonometric function. Let $u(x) = x$ and $v'(x) = \cos x$. It is helpful to set up a table as shown here to organize the functions.

$u(x) = x$	$v'(x) = \cos x$
$u'(x) = 1$	$v(x) = \sin x$

We can then use the integration by parts formula:

$$\int x\cos x dx = x\sin x - \int 1 \cdot \sin x dx$$

The integral on the right can be evaluated easily, and we have:

$$\int x\cos x dx = x\sin x + \cos x + c$$

Example 2

Evaluate $\int xe^x dx$.

Solution: This integrand involves a product of a polynomial and an exponential function. Let $u(x) = x$ and $v'(x) = e^x$. Set up a table to organize the functions.

$u(x) = x$	$v'(x) = e^x$
$u'(x) = 1$	$v(x) = e^x$

We can then use the integration by parts formula:

$$\int xe^x dx = xe^x - \int e^x dx$$

The integral on the right can be evaluated easily, and we have:

$$\int xe^x dx = xe^x - e^x + c$$

Example 3

Evaluate $\int \ln x dx$.

Solution: This integrand is tricky, in that there is only one function visible. Keep in mind that there is always a "1" in an integrand: $\int \ln x dx = \int 1 \ln x dx$. One of the functions, either 1 or $\ln x$, will have to be differentiated; the other will be integrated. We do not know how to integrate $\ln x$ yet, so that will have to be the function to differentiate. The other function will be the one to integrate:

$u(x) = \ln x$	$v'(x) = 1$
$u'(x) = \frac{1}{x}$	$v(x) = x$

We can then use the integration by parts formula:

$$\int \ln x dx = x \ln x - \int x \cdot \tfrac{1}{x} dx$$

The integral on the right can be evaluated easily, and we have:

$$\int \ln x dx = x \ln x - \int 1 dx = x \ln x - x + c$$

Example 4

Evaluate the integral $\int x \tan^{-1} x dx$.

Solution: This integrand involves the product of a polynomial and an inverse trigonometric function. In this case, we do not have a formula for the integral of the arctangent function, but we do know how to differentiate it. Let $u(x) = \tan^{-1} x$ and $v'(x) = x$. It is helpful to set up a table as shown here to organize the functions.

$u(x) = \tan^{-1}x$	$v'(x) = x$
$u'(x) = \frac{1}{1+x^2}$	$v(x) = \frac{1}{2}x^2$

We can then use the integration by parts formula:

$$\int x\tan^{-1} x\,dx = \tfrac{1}{2}x^2 \tan^{-1} x - \int \tfrac{1}{2}x^2\left(\tfrac{1}{1+x^2}\right)dx = \tfrac{1}{2}x^2 \tan^{-1} x - \tfrac{1}{2}\int\left(\tfrac{x^2}{1+x^2}\right)dx$$

The integral on the right can be evaluated by simplifying the integrand. The integrand can be simplified by division, or by adding and subtracting 1 to the numerator:

$$\frac{x^2}{1+x^2} = \frac{x^2+1-1}{1+x^2} = \frac{x^2+1}{1+x^2} - \frac{1}{1+x^2} = 1 - \frac{1}{1+x^2}$$

Our integral now becomes:

$$\int x\tan^{-1} x\,dx = \tfrac{1}{2}x^2 \tan^{-1} x - \tfrac{1}{2}\int\left(1 - \tfrac{1}{1+x^2}\right)dx$$
$$= \tfrac{1}{2}x^2 \tan^{-1} x - \tfrac{1}{2}\left(x - \tan^{-1} x\right) + c$$

Lesson 15-3 Review

Evaluate the following integrals:

1. $\int x^2 \ln x\,dx$
2. $\int x \sec^2 x\,dx$

Lesson 15-4: Integrating Rational Functions

Some rational functions can be integrated using the substitution technique. This technique applies if the numerator of the rational function is equal to the derivative of the denominator of the rational function. In this situation, setting $u(x)$ equal to the denominator of the rational function effectively wipes out the numerator, and the integral that remains is $\int \frac{1}{u}du$. If the numerator is *not* equal to the derivative of the denominator, the substitution approach will not help, but there is another method to use.

The most useful method for integrating rational functions makes use of the **partial fraction decomposition** of a rational function. Try this technique when the denominator of the rational function can be factored. The motivation behind partial fraction decomposition involves reversing the process of adding fractions. Any polynomial $P(x)$ can be factored as a product of linear factors of the form $(ax + b)$ and irreducible quadratic factors of the form $ax^2 + bx + c$. An **irreducible quadratic factor** is a quadratic expression that has no real zeros. The expression $x^2 + 1$ is an example of an irreducible quadratic factor. To decompose a rational function $\frac{P(x)}{Q(x)}$, first factor the polynomial in the denominator, $Q(x)$, completely. Write the rational function $\frac{P(x)}{Q(x)}$ as a sum of fractions of the form $\frac{A}{(ax+b)^i}$ and $\frac{Ax+B}{(ax^2+bx+c)^i}$. The fractions to include depend on the nature of the factors of $Q(x)$.

Suppose the factors of $Q(x)$ are all distinct and linear: $(a_1x + b_1), (a_2x + b_2), \ldots, (a_nx + b_n)$.

Write $\frac{P(x)}{Q(x)} = \frac{A_1}{(a_1x+b_1)} + \frac{A_2}{(a_2x+b_2)} + \cdots \frac{A_n}{(a_nx+b_n)}$ and solve for the constants $A_1, A_2, \ldots, A_n$.

Then: $\int \frac{P(x)}{Q(x)} dx = \int \left(\frac{A_1}{(a_1x+b_1)} + \frac{A_2}{(a_2x+b_2)} + \cdots \frac{A_n}{(a_nx+b_n)} \right) dx$.

We can integrate each term using substitution:

$$\int \frac{P(x)}{Q(x)} dx = \frac{A_1}{a_1} \ln|a_1x+b_1| + \frac{A_2}{a_2} \ln|a_2x+b_2| + \ldots + \frac{A_n}{a_n} \ln|a_nx+b_n| + c.$$

Example 1

Evaluate $\int \frac{1}{(x-2)(x+3)} dx$.

Solution: Decompose the rational function into individual fractions:

$$\frac{1}{(x-2)(x+3)} = \frac{A_1}{(x-2)} + \frac{A_2}{(x+3)}$$

Solve for the constants A_1 and A_2:

$$\frac{1}{(x-2)(x+3)} = \frac{A_1}{(x-2)} + \frac{A_2}{(x+3)}$$

Multiply both sides of the equation by $(x - 2)(x + 3)$

$$1 = A_1(x + 3) + A_2(x - 2)$$

Substitute $x = -3$ into this equation and solve for A_2

$$1 = A_1(-3 + 3) + A_2(-3 - 2)$$

$$A_2 = -\tfrac{1}{5}$$

Substitute $x = 2$ into the equation $1 = A_1(x + 3) + A_2(x - 2)$ and solve for A_1

$$1 = A_1(2 + 3) + A_2(2 - 2)$$

$$A_1 = \tfrac{1}{5}$$

So $\frac{1}{(x-2)(x+3)} = \frac{\frac{1}{5}}{(x-2)} - \frac{\frac{1}{5}}{(x+3)}$, and we have:

$$\int \frac{1}{(x-2)(x+3)}dx = \tfrac{1}{5}\int\left(\frac{1}{(x-2)} - \frac{1}{(x+3)}\right)dx = \tfrac{1}{5}\left(\ln|x-2| - \ln|x+3|\right) + c$$

Suppose the factors of $Q(x)$ are all linear, but some factors repeat. Suppose the factor $(a_1x + b_1)$ is repeated r times. Then the linear factor $(a_1x + b_1)$ will be repeated r times, in the form:

$$\frac{A_1}{(a_1x+b_1)} + \frac{A_2}{(a_2x+b_2)^2} + \dots \frac{A_r}{(a_nx+b_n)^r}$$

We would need to solve for the constants A_1, A_2, ..., A_r. The non-repeating factors are handled as before. Substitute the partial fraction decomposition of $\frac{P(x)}{Q(x)}$ and integrate each term using substitution.

Example 2

Find the partial fraction decomposition of the rational function $\frac{x^2+1}{x(x-1)^3}$.

Solution: There is one linear factor that repeats 3 times, so we have:

$$\frac{x^2+1}{x(x-1)^3}=\frac{A_1}{x}+\frac{A_2}{(x-1)}+\frac{A_3}{(x-1)^2}+\frac{A_4}{(x-1)^3}$$

Solve for the constants A_1, A_2, A_3, and A_4

$$\frac{x^2+1}{x(x-1)^3}=\frac{A_1}{x}+\frac{A_2}{(x-1)}+\frac{A_3}{(x-1)^2}+\frac{A_4}{(x-1)^3}$$

Multiply both sides of this equation by $x(x-1)^3$

$$x^2+1=A_1(x-1)^3+A_2x(x-1)^2+A_3x(x-1)+A_4x$$

Evaluate this equation at $x = 0$ to solve for A_1 and at $x = 1$ to solve for A_2

$$A_1=-1 \text{ and } A_4=2$$

Evaluate this equation for two other values of x to create a system of equations in terms of A_2 and A_3. Use $x = 2$ and $x = 3$.

$$x^2+1=-1(x-1)^3+A_2x(x-1)^2+A_3x(x-1)+2x$$

$$5=-1+2A_2+2A_3+4$$

$$10=-8+12A_2+6A_3+6$$

Solve this system of equations

$$\begin{cases}2A_2+2A_3=2\\12A_2+6A_3=12\end{cases}$$

$A_2=1$ and $A_3=0$

The partial fraction decomposition is $\frac{x^2+1}{x(x-1)^3}=-\frac{1}{x}+\frac{1}{(x-1)}+\frac{2}{(x-1)^3}$.

If the factors of $Q(x)$ include a quadratic expression that has no real zeros, then the numerator of the fraction that incorporates the quadratic expression must be of the form $Bx + C$. Repeating factors are handled in the same way we discussed earlier. For example, the partial fraction decomposition of $\frac{1}{x(x^2+1)}$ is: $\frac{1}{x(x^2+1)}=\frac{A_1}{x}+\frac{B_1x+C_1}{(x^2+1)}$.

The partial fraction decomposition of

$$\frac{1}{x^3\left(x^2+1\right)^2} \text{ is: } \frac{1}{x^3\left(x^2+1\right)^2}=\frac{A_1}{x}+\frac{A_2}{x^2}+\frac{A_3}{x^3}+\frac{B_1x+C_1}{\left(x^2+1\right)}+\frac{B_2x+C_2}{\left(x^2+1\right)^2}$$

Once a rational function is decomposed, each fraction can be integrated using substitution or some other technique. Partial fraction decomposition should be used if a simple substitution does not work.

Lesson 15-4 Review

1. Evaluate $\int \frac{x}{x^2+2x-3}dx$.

Lesson 15-5: A Strategy for Integration

The skills necessary for integration are based on the techniques used for differentiation. You must be very familiar with the integrals of the elementary functions that were discussed in Lesson 15-1. Every other technique involves rewriting the integrand so that it resembles one of these functions. Using the substitution technique requires the ability to differentiate using the chain rule. Integration by parts requires you to understand each piece in the product rule. The following strategy for integration may help you integrate some functions. As your integration skills improve, it is important to recognize your limitations. There are some functions that do not have an anti-derivative that can be written compactly.

When trying to evaluate an integral, the first technique to apply is to see if you can simplify the integrand. Multiplying polynomials or using a trigonometric identity may turn a hard problem into an easy problem. For example, $\int \frac{e^x+1}{e^x}dx$ can be evaluated by breaking up the fraction: $\int \frac{e^x+1}{e^x}dx=\int\left(1+e^{-x}\right)dx$. Evaluating $\int \frac{\tan x}{\sin x \sec x}dx$ can be simplified by writing the functions in terms of sine and cosine functions and simplifying:

$$\int \frac{\tan x}{\sin x \sec x}dx=\int \frac{\sin x}{\cos x}\cdot\frac{1}{\sin x}\cos x dx=\int 1dx.$$

If simplifying the integrand does not help, look for an obvious substitution. Do not try to make your substitutions too involved. Keep it simple, but not too simple. For example, $\int \tan x dx$ can be evaluated by writing the function in terms of sines and cosines, and then using the substitution $u = \cos x$:

$$\int \tan x dx = \int \frac{\sin x}{\cos x} dx = -\int \frac{1}{u} du = -\ln|u| + c = -\ln|\cos x| + c$$

If the integrand is a rational function, look for an obvious substitution.

If there is no obvious substitution, look carefully at the integrand. If the integrand involves a product of different types of functions, such as polynomials and trigonometric functions, exponential, and trigonometric functions, or polynomials and logarithm functions, try integration by parts. If the integrand is a rational function, try rewriting the rational function in terms of its partial fractions.

If you are still having trouble, try a more complicated substitution method. Try a different trigonometric identity. Try a different simplification of the integrand. Try to relate the problem to one that you have previously solved. Try a table of integrals. Try to combine methods, such as substitution and then parts. Try taking a break from the problem. Of course, you may be trying to find an anti-derivative that cannot be written in terms of elementary functions, as is the case with $\int e^{x^2} dx$. Evaluating some integrals requires *more* than calculus.

Example 1

Evaluate $\int e^{\sqrt{x}} dx$.

Solution: Try the substitution $u = \sqrt{x}$, $du = \frac{1}{2\sqrt{x}} dx$. There is a $\sqrt{x}$ in the denominator of the differential equation that will not be absorbed when we make the substitution. In this case, use the equation $u = \sqrt{x}$ to get rid of it:

$$du = \frac{1}{2\sqrt{x}} dx$$

$$2\sqrt{x} du = dx$$

$$2u du = dx$$

Substitute into the integral:

$$\int e^{\sqrt{x}} dx$$

$$\int 2ue^{u} du$$

This integral can now be evaluated using integration by parts, as we saw in Lesson 15-3: $\int 2ue^u du = 2(ue^u - e^u) + c$. Finally, substitute back into the original variable: $\int e^{\sqrt{x}} dx = 2(\sqrt{x}e^{\sqrt{x}} - e^{\sqrt{x}}) + c$

Lesson 15-5 Review

Evaluate the following integrals:

1. $\int \frac{e^{2x}}{1+e^x} dx$

2. $\int \frac{x}{(x+2)^2} dx$

Lesson 15-6: Area and the Riemann Sum

We will begin this lesson with a discussion of a method to measure the area under a curve. This method involves calculus, but it does not involve derivatives. Not yet, anyway. We will measure the area under a curve by dividing the region under the curve into little rectangles and measuring their area. The area of a rectangle is the product of the length of the rectangle and the height of the rectangle. Suppose that $f(x)$ is a continuous function on a closed interval $[a, b]$, as shown in Figure 15.1. Divide $[a, b]$ into n partitions of equal length. If we denote the length of each partition by Δx, then the length of each partition can be found by the equation $\Delta x = \frac{b-a}{n}$.

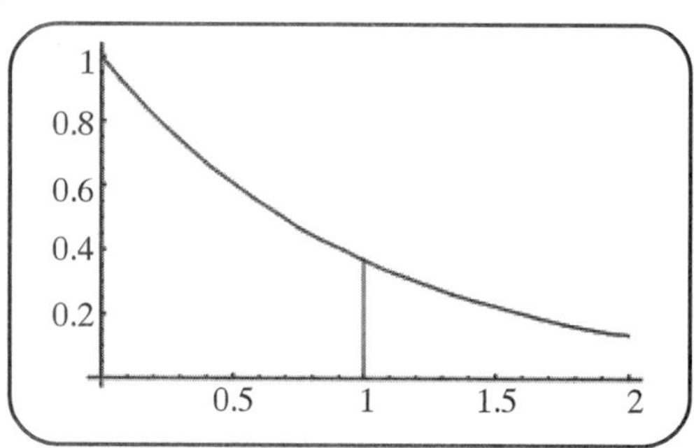

Figure 15.1.

Let $x_0, x_1, \ldots, x_n$ be the endpoints of the subdivisions. We can construct little rectangles whose base is one of the partitions of the interval and whose height extends up to the function value at a point in the partition.

Choosing different partition points to evaluate the function will give different values for the area of the rectangle. If we systematically choose the left endpoint of the partition to evaluate the function and create the rectangle, we will measure the area of the region shown in Figure 15.2. This estimate for the area is called the left endpoint estimate. The equation for calculating the area is:

$$A_L = f(x_0)\Delta x + f(x_1)\Delta x + \ldots + f(x_{n-1})\Delta x$$

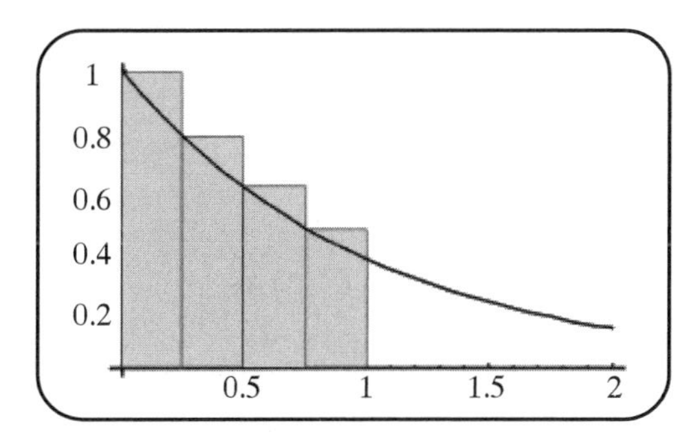

Figure 15.2.

We can write this more compactly using summation notation, or sigma notation:

$$A_L = \sum_{i=0}^{n-1} f(x_i)\Delta x$$

The symbol Σ is a capital sigma, which is the Greek letter for "S," and is used to represent a sum. In this compact notation, we are instructed to add together terms of the form $f(x_i)\Delta x$. The variable i serves as a counter. The term $i = 0$ at the bottom indicates the starting point of the counter, and the term $n - 1$ at the top indicates the stopping point. The sums that appear in this lesson are called **Riemann sums.**

If we systematically choose the right endpoint of the partition to evaluate the function and create the rectangle, we will measure the area of the region shown in Figure 15.3 on page 268. This estimate for the area is called the right endpoint estimate. The equation for calculating the area is:

$$A_R = f(x_1)\Delta x + f(x_2)\Delta x + \ldots + f(x_n)\Delta x$$

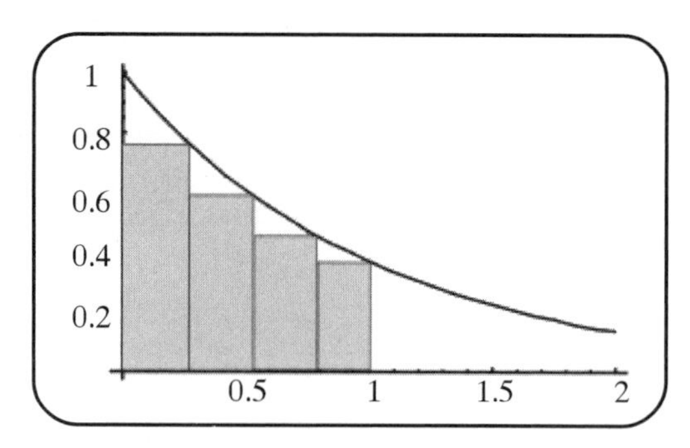

Figure 15.3.

We can write this more compactly using summation notation, or sigma notation:

$$A_R = \sum_{i=1}^{n} f(x_i)\Delta x$$

The difference between the sigma notation equations for A_L and A_R is in the starting and stopping value of i.

If we systematically choose the midpoint of the partition to evaluate the function and create the rectangle, we will measure the area of the region shown in Figure 15.4. This estimate for the area is called the midpoint estimate. The equation for calculating the area is:

$$A_M = f(x_1^*)\Delta x + f(x_2^*)\Delta x + \ldots + f(x_n^*)\Delta x$$

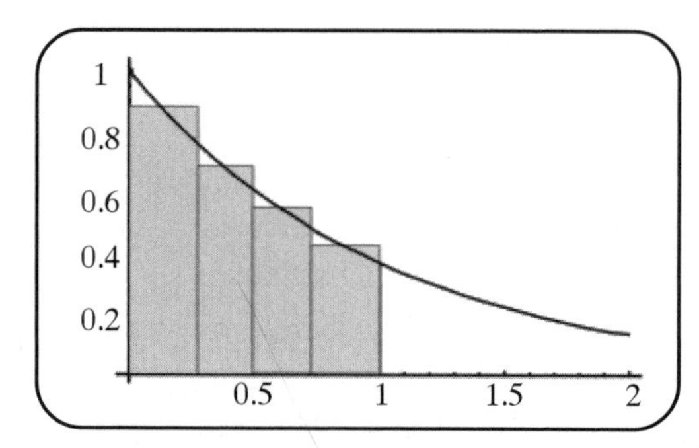

Figure 15.4.

We can write this more compactly using summation notation, or sigma notation:

$$A_M = \sum_{i=1}^{n} f(x_i^*)\Delta x$$

The left and right endpoint methods are usually the least accurate. The average of these two values is called the trapezoidal estimate for the area. The geometric interpretation for the trapezoidal estimate is shown in Figure 15.5. In essence, averaging the left and right endpoint estimates has the same numerical value as adding up the areas of the little trapezoids shown. The equation for the trapezoidal estimate is:

$$A_T = \frac{A_L + A_R}{2}$$

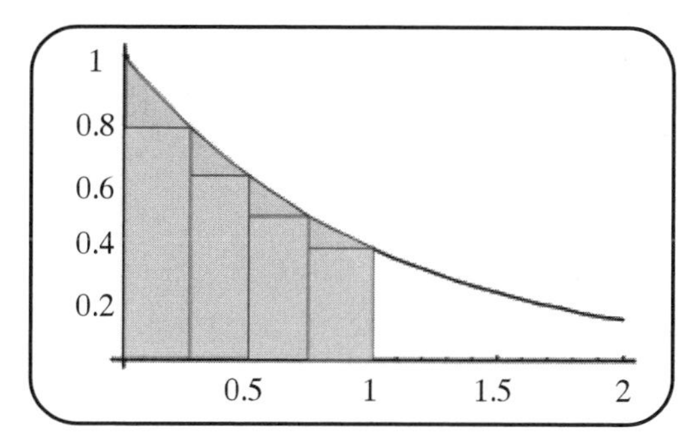

Figure 15.5.

The trapezoidal method is generally more accurate than either the left or the right endpoint methods. The estimate using the midpoints of the partitions is generally more accurate than the trapezoidal method. Taking a weighted average of the midpoint and trapezoidal estimates improves the accuracy even more. This method is called **Simpson's Method.** The formula for Simpson's Method can be written:

$$A_S = \frac{A_T + 2A_M}{3}$$

Example 1

Estimate the area under the curve $f(x) = e^x$ and above the x-axis, between $x = 0$ and $x = 2$. Break the interval [0, 2] into four partitions, and give the left endpoint, the right endpoint, the trapezoidal, the midpoint, and Simpson's estimates for this area.

Solution: The graph of $f(x) = e^x$ over the interval [0, 2] is shown in Figure 15.6 on page 270.

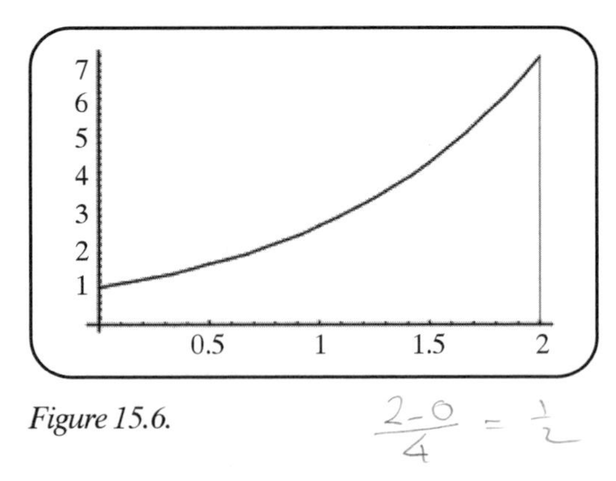

Figure 15.6.

If we break the interval [0, 2] into four partitions, then each partition will have length:

$$\Delta x = \frac{b-a}{n} = \frac{2-0}{4} = \frac{1}{2}$$

The endpoints are the points $x_0 = 0$, $x_1 = \frac{1}{2} = 0.5$, $x_2 = \frac{2}{2} = 1$, $x_3 = \frac{3}{2} = 1.5$, $x_4 = \frac{4}{2} = 2$.

The left endpoint method will use $x_0 = 0$, $x_1 = \frac{1}{2} = 0.5$, $x_2 = \frac{2}{2} = 1$, and $x_3 = \frac{3}{2} = 1.5$:

$$A_L = f(x_0)\Delta x + f(x_1)\Delta x + f(x_2)\Delta x + f(x_3)\Delta x$$

$$A_L = (e^0)\tfrac{1}{2} + (e^{0.5})\tfrac{1}{2} + (e^1)\tfrac{1}{2} + (e^{1.5})\tfrac{1}{2} \approx 4.924$$

The right endpoint method will use $x_1 = \frac{1}{2} = 0.5$, $x_2 = \frac{2}{2} = 1$, $x_3 = \frac{3}{2} = 1.5$, $x_4 = \frac{4}{2} = 2$:

$$A_R = f(x_1)\Delta x + f(x_2)\Delta x + f(x_3)\Delta x + f(x_4)\Delta x$$

$$A_R = (e^{0.5})\tfrac{1}{2} + (e^1)\tfrac{1}{2} + (e^{1.5})\tfrac{1}{2} + (e^2)\tfrac{1}{2} \approx 8.119$$

The trapezoidal method is $A_T = \frac{A_L + A_R}{2} = \frac{4.924 + 8.119}{2} = 6.522$.

The midpoint method will use $x_1^* = 0.25$, $x_2^* = 0.75$, $x_3^* = 1.25$, $x_4^* = 1.75$:

$$A_M = f(x_1^*)\Delta x + f(x_2^*)\Delta x + f(x_3^*)\Delta x + f(x_4^*)\Delta x$$

$$A_M = (e^{0.25})\tfrac{1}{2} + (e^{0.75})\tfrac{1}{2} + (e^{1.25})\tfrac{1}{2} + (e^{1.75})\tfrac{1}{2} \approx 6.323$$

Simpson's Method gives: $A_S = \frac{A_T + 2A_M}{3} = \frac{6.522 + 2(6.323)}{3} = 6.389$.

The actual value for the area of the region in Example 1 is 6.38906. Simpson's Method gives a remarkably accurate estimate, even though we effectively broke this area up into only four rectangles!

The shape of a curve will determine whether our approximations are overestimates or underestimates. If a function is increasing concave up, as in the case with $f(x) = e^x$, then the left endpoint method will be an underestimate, and the right endpoint method will be an overestimate. The trapezoidal method will be an overestimate, and the midpoint method will be an underestimate. We can analyze the different methods for different shapes. The results are summarized in the following table.

Shape	Left	Right	Trapezoidal	Midpoint
↑ ∪	Under	Over	Over	Under
↑ ∩	Under	Over	Under	Over
↓ ∪	Over	Under	Over	Under
↓ ∩	Over	Under	Under	Over

The methods used to calculate the area of a region do not involve sophisticated mathematics. We divided an interval up into n partitions and evaluated the function at the endpoints of the partitions. After some multiplication and addition, we averaged our answers and obtained a fairly accurate estimate. There was no calculus involved, but that is about to change.

Our estimate for the area in Example 1 was fairly accurate using only four partitions. The more partitions used, the more accurate the estimate. We could have used use 10 partitions, or 1,000 partitions, or even 1,000,000! As we saw with compound interest, we can push our area estimates to the limit and use infinitely many partitions! If the number of partitions approached infinity, the error in our estimate for the area would get smaller, and our estimates for the area would all approach the actual value for the area. This is, of course, assuming that these limits exist! Incorporating a limit into these calculations gives yet another example of how calculus can be used to analyze a function. This application involves adding little bits of area together to calculate the area under a curve.

As the number of partitions heads towards infinity, we will require that all of our estimates (the left and right endpoint, the midpoint, the trapezoidal, and Simpson's Method) all converge, or head towards, the same value. In this case, it will not matter which method we use when we take the limit.

With every new mathematical idea comes new notation. We are actually going to recycle one of our symbols. As the number of partitions, n, approaches infinity, we will let our symbol for summation, Σ, become a new symbol that also resembles the letter "S": $\int$. We write:

$$\lim_{n\to\infty}\sum_{i=0}^{n-1} f(x_i)\Delta x = \int_a^b f(x)dx$$

Although we are reusing the symbol for the integral, $\int$, there are some differences in the notation. In the expression $\int_a^b f(x)dx$, which we refer to as a **definite integral**, the numbers a and b represent the endpoints of the interval.

If a function is positive, then the actual area under the curve and the value of the definite integral are the same. If a function is negative on part of the interval $[a, b]$, then the definite integral cannot be interpreted as the area under a curve, but it can still be approximated by its Riemann sum:

$$\int_a^b f(x)dx = \lim_{n\to\infty}\sum_{i=0}^{n-1} f(x_i)\Delta x$$

Lesson 15-6 Review

1. Evaluate $\int_0^4 (x^2+1)dx$ by breaking the interval [0, 4] into four partitions, and give the left endpoint, the right endpoint, the trapezoidal, the midpoint, and Simpson's estimates for this area.

Lesson 15-7: The Fundamental Theorem of Calculus

The fact that we reused the integral in the last lesson should raise some eyebrows. With the abundance of symbols available, we could have found some other symbol to represent a sum if we really wanted to. There is an amazing relationship between an anti-derivative and a definite integral, and the **Fundamental Theorem of Calculus** is the mathematical result that makes their connection clear.

The process of finding an anti-derivative of a function involves un-differentiating a function. The *derivative* of a function involves taking the limit of the difference quotient, or the limit of a quotient of differences. The formula for a *definite integral* involves taking the limit of sums of products. The derivative (quotient of differences) and the definite integral (sums of products) are like opposites and may remind you of the inverse of a process.

There are actually two parts to the Fundamental Theorem of Calculus. The first part can be stated as follows:

The Fundamental Theorem of Calculus, Part I

If $f(x)$ is continuous on a closed interval $[a, b]$, and x is between a and b, then the function defined by $g(x) = \int_a^x f(x)dx$ is continuous on $[a, b]$ and differentiable on (a, b), and satisfies the equation $g'(x) = f(x)$.

This part of the Fundamental Theorem of Calculus basically says that if you integrate a function and then take its derivative, you will get back what you started with. In other words, a definite integral can be undone by differentiation.

The second part of the Fundamental Theorem of Calculus is probably the most useful part, as far as calculations are concerned. The second part can be stated as follows:

The Fundamental Theorem of Calculus, Part II

If $f(x)$ is continuous on a closed interval $[a, b]$, and if $F(x)$ is any anti-derivative of $f(x)$, then $\int_a^b f(x)dx = F(b) - F(a)$.

This part of the Fundamental Theorem of Calculus gives us a shortcut to evaluating definite integrals. Instead of computing the limit of the Riemann sums, we can compute the definite integral of a function by evaluating *any* anti-derivative of the function at the upper and lower endpoints of the interval and subtracting. There is no need to partition the interval and estimate the integral by the left endpoint, right endpoint, midpoint, and trapezoidal methods. Of course, this assumes that the function $f(x)$ has an anti-derivative that can be found! For those functions that have no simple anti-derivative, such as $f(x) = e^{x^2}$, you must return to Riemann sums to estimate the definite integral.

Example 1

Use the Fundamental Theorem of Calculus to evaluate the following definite integrals:

a. $\int_0^3 (x^2+1)dx$

b. $\int_0^\pi \sin x dx$

c. $\int_1^4 \ln x dx$

d. $\int_0^2 xe^{x^2} dx$

Solution: First find an anti-derivative (using whatever integration techniques you need), and then evaluate the anti-derivative at the upper and lower limits and subtract. Because you can use *any* anti-derivative, we will omit the integration constant.

a. $\int_0^3 (x^2+1)dx$: $F(x) = \int (x^2+1)dx = \frac{1}{3}x^3 + x$,

$$\int_0^3 (x^2+1)dx = F(3) - F(0) = \left(\tfrac{1}{3}(3)^3 + 3\right) - \left(\tfrac{1}{3}(0)^3 + 0\right) = 12$$

b. $\int_0^\pi \sin x dx$: $F(x) = \int \sin x dx = -\cos x$,

$$\int_0^\pi \sin x dx = F(\pi) - F(0) = (-\cos\pi) - (-\cos 0) = 1 - (-1) = 2$$

c. $\int_1^4 \ln x dx$: $F(x) = \int \ln x dx = x\ln x - x$,

$$\int_1^4 \ln x dx = F(4) - F(1) = (4\ln 4 - 4) - (1\ln 1 - 1) = 4\ln 4 - 3$$

d. $\int_0^2 xe^{x^2} dx$: $F(x) = \int xe^{x^2} dx = \frac{1}{2}e^{x^2}$,

$$\int_0^2 xe^{x^2} dx = F(2) - F(0) = \tfrac{1}{2}e^4 - \tfrac{1}{2}e^0 = \tfrac{1}{2}e^4 - \tfrac{1}{2}$$

The Fundamental Theorem of Calculus makes the connection between the definite integral and the derivative. It can be written as:

$$\int_a^b f'(x)dx = f(b) - f(a)$$

From this equation, we can see that the definite integral of the rate of change of a function gives the total change of the function over the interval $[a, b]$. In other words, adding up all the little bits of change over an interval yields the total overall change of the function over the interval.

There are many applications of the integral and the Fundamental Theorem of Calculus. For example, we can use the Fundamental Theorem of Calculus to reconstruct a function from its derivative. In Chapters 13 and 14 we caught a glimpse of how to reconstruct the graph of a function from the graph of its derivative. The Fundamental Theorem of Calculus can be used to draw a precise graph of the function by measuring the area under the graph of its derivative and using the equation:

$$f(b) = f(a) + \int_a^b f'(x)dx$$

Integration is a rich topic, and in this Chapter I have just barely scratched its surface. I hope this introduction to calculus has left you curious about the subject matter and wanting to learn more. The subject has evolved over the years, and although technological advances may have outdated some of the applications, it is far too useful to ever become extinct.

Lesson 15-7 Review

Use the Fundamental Theorem of Calculus to evaluate the following definite integrals:

1. $\int_3^5 \frac{(x+1)}{x^2+2x-8}dx$

2. $\int_0^\pi x\sin x dx$

3. $\int_1^4 \frac{\ln x}{x} dx$

Answer Key

Lesson 15-1 Review

1. $\int\left(3x^4+\sinh x\right)dx=\frac{3}{5}x^5+\cosh x+c$

2. $\int\left(\frac{3}{\sqrt{x}}+2\sec^2 x\right)dx=\int\left(3x^{-1/2}+2\sec^2 x\right)dx=\frac{3}{1/2}x^{1/2}+2\tan x+c$

   $=6x^{1/2}+2\tan x+c$

3. $\int\left(2e^x+\sec x\tan x\right)dx=2e^x+\sec x+c$

4. $\int\left(\frac{3}{x^5}-\frac{4}{x}\right)dx=\int\left(3x^{-5}-4x^{-1}\right)dx=\frac{3}{-4}x^{-4}-4\ln|x|+c$

Lesson 15-2 Review

1. $\int\sin(2x+3)dx$: $u=2x+3$, $du=2dx$:

   $\int\sin(2x+3)dx=\frac{1}{2}\int\sin u\,du=-\frac{1}{2}\cos u+c=-\frac{1}{2}\cos(2x+3)+c$

2. $\int\frac{(1+\sqrt{x})^9}{\sqrt{x}}dx$: $u=1+\sqrt{x}$, $du=\frac{1}{2\sqrt{x}}dx$:

   $\int\frac{(1+\sqrt{x})^9}{\sqrt{x}}dx=2\int u^9du=2\left(\frac{u^{10}}{10}\right)+c=\frac{1}{5}\left(1+\sqrt{x}\right)^{10}+c$

3. $\int x\sqrt{x-1}dx$: $u=x-1$, $du=dx$, $x=u+1$:

   $\int x\sqrt{x-1}dx=\int(u+1)\sqrt{u}du$

   $=\int\left(u^{3/2}+u^{1/2}\right)du=\frac{u^{5/2}}{5/2}+\frac{u^{3/2}}{3/2}+c=\frac{2}{5}(x-1)^{5/2}+\frac{2}{3}(x-1)^{3/2}+c$

Lesson 15-3 Review

1. $\int x^2\ln x\,dx$:

$u(x)=\ln x$	$v'(x)=x^2$
$u'(x)=\frac{1}{x}$	$v(x)=\frac{1}{3}x^3$

   $\int x^2\ln x\,dx=\frac{1}{3}x^3\ln x-\int\left(\frac{1}{x}\right)\left(\frac{1}{3}x^3\right)dx=\frac{1}{3}x^3\ln x-\frac{1}{3}\int x^2dx=\frac{1}{3}x^3\ln x-\frac{1}{9}x^3+c$

2. $\int x\sec^2 x dx$:

$u(x)=x$	$v'(x)=\sec^2 x$
$u'(x)=1$	$v(x)=\tan x$

$$\int x\sec^2 x dx = x\tan x - \int 1\cdot \tan x dx = x\tan x + \ln|\cos x| + c$$

Lesson 15-4 Review

1. $\int \frac{x}{x^2+2x-3}dx$: $\frac{x}{x^2+2x-3} = \frac{x}{(x+3)(x-1)} = \frac{A_1}{(x+3)} + \frac{A_2}{(x-1)} = \frac{1}{4}\left(\frac{3}{(x+3)} + \frac{1}{(x-1)}\right)$

$$\int \frac{x}{x^2+2x-3}dx = \frac{1}{4}\int \left(\frac{3}{x+3} + \frac{1}{x-1}\right)dx = \frac{1}{4}\left(3\ln|x+3| + \ln|x-1|\right) + c$$

Lesson 15-5 Review

1. Let $u = 1+e^x$, $du = e^x dx$, and $(u-1) = e^x$:

$$\int \frac{e^{2x}}{1+e^x}dx = \int \frac{e^x \cdot e^x}{1+e^x}dx = \int \frac{(u-1)}{u}du$$
$$= \int \left(1-\tfrac{1}{u}\right)du = u - \ln|u| + c = \left(e^x+1\right) - \ln\left|1+e^x\right| + c$$

2. Let $u = x+2$, $du = dx$, and $(u-2) = x$:

$$\int \frac{x}{(x+2)^2}dx = \int \frac{(u-2)}{u^2}du = \int \left(\tfrac{1}{u} - 2u^{-2}\right)du = \ln|u| + \tfrac{2}{u} + c = \ln|x+2| + \tfrac{2}{x+2} + c$$

Lesson 15-6 Review

1. If we break the interval [0, 4] into four partitions, then each partition will have length:

   $\Delta x = \frac{b-a}{n} = \frac{4-0}{4} = 1$.

   The endpoints are the points $x_0 = 0, x_1 = 1, x_2 = 2, x_1 = 3, x_4 = 4$.

   The left endpoint method will use $x_0 = 0, x_1 = 1, x_2 = 2$, and $x_1 = 3$:

   $A_L = (1)\cdot 1 + (2)\cdot 1 + (5)\cdot 1 + (10)\cdot 1 = 18$

   The right endpoint method will use $x_1 = 1, x_2 = 2, x_1 = 3, x_4 = 4$:

   $A_R = (2)\cdot 1 + (5)\cdot 1 + (10)\cdot 1 + (17)\cdot 1 = 34$

The trapezoidal method is $A_T = \frac{A_L + A_R}{2} = \frac{18+34}{2} = 26.$

The midpoint method will use $x_1^* = 0.5,\ x_2^* = 1.5,\ x_3^* = 2.5,\ x_4^* = 3.5$:

$A_M = (1.25)\cdot 1 + (3.25)\cdot 1 + (7.25)\cdot 1 + (13.25)\cdot 1 = 25.$

Simpson's Method gives: $A_S = \frac{A_T + 2A_M}{3} = \frac{26+2(25)}{3} = 25.33.$

Lesson 15-7 Review

1. $\int_3^5 \frac{(x+1)}{x^2+2x-8}dx$: $\frac{x+1}{x^2+2x-8} = \frac{x+1}{(x+4)(x-2)} = \frac{A_1}{(x+4)} + \frac{A_2}{(x-2)} = \frac{1}{2}\left(\frac{1}{(x+4)} + \frac{1}{(x-2)}\right)$

$F(x) = \int \frac{(x+1)}{x^2+2x-8}dx = \frac{1}{2}\int\left(\frac{1}{x+4} + \frac{1}{x-2}\right)dx = \frac{1}{2}\left(\ln|x+4| + \ln|x-2|\right)$

$\int_3^5 \frac{(x+1)}{x^2+2x-8}dx = F(5) - F(3) = \frac{1}{2}\left(\ln|5+4| + \ln|5-2|\right) - \frac{1}{2}\left(\ln|3+4| + \ln|3-2|\right)$

$\int_3^5 \frac{(x+1)}{x^2+2x-8}dx = \frac{1}{2}(\ln 9 + \ln 3) - \frac{1}{2}(\ln 7 + \ln 1) = \frac{1}{2}\ln\left(\frac{9\cdot 3}{7}\right) = \frac{1}{2}\ln\left(\frac{27}{7}\right)$

2. $\int_0^\pi x\sin x dx$: $F(x) = \int x\sin x dx = -x\cos x + \sin x$,

$\int_0^\pi x\sin x dx = F(\pi) - F(0) = (-\pi\cos\pi + \sin\pi) - (-0\cos 0 + \sin 0) = \pi$

3. $\int_1^4 \frac{\ln x}{x}dx$: $F(x) = \int \frac{\ln x}{x}dx = \frac{1}{2}(\ln x)^2$,

$\int_1^4 \frac{\ln x}{x}dx = F(4) - F(1) = \frac{1}{2}(\ln 4)^2 - \frac{1}{2}(\ln 1)^2 = \frac{1}{2}(\ln 4)^2$

Index

D

E

F

U

X

Y

Z

About the Author

DENISE SZECSEI earned Bachelor of Science degrees in physics, chemistry, and mathematics from the University of Redlands, and she was greatly influenced by the educational environment cultivated through the Johnston Center for Integrative Studies. After graduating from the University of Redlands, she served as a technical instructor in the U.S. Navy. After completing her military service, she earned a Ph.D in mathematics from the Florida State University. She has been teaching since 1985.